Cellular Computing

Series in Systems Biology
Edited by Dennis Shasha, New York University

Editorial Board
Michael Ashburner, University of Cambridge
Amos Bairoch, Swiss Institute of Bioinformatics
David Botstein, Princeton University
Charles Cantor, Sequenom, Inc.
Leroy Hood, Institute for Systems Biology
Minoru Kanehisa, Kyoto University
Raju Kucherlapati, Harvard Medical School

Systems Biology describes the discipline that seeks to understand biological phenomena on a large scale: the association of gene with function, the detailed modeling of the interaction among proteins and metabolites, and the function of cells. Systems Biology has wide-ranging application, as it is informed by several underlying disciplines, including biology, computer science, mathematics, physics, chemistry, and the social sciences. The goal of the series is to help practitioners and researchers understand the ideas and technologies underlying Systems Biology. The series volumes will combine biological insight with principles and methods of computational data analysis.

Cellular Computing
Martyn Amos, University of Exeter

Cellular Computing

Edited by

Martyn Amos

OXFORD
UNIVERSITY PRESS

2004

OXFORD

UNIVERSITY PRESS

Oxford   New York
Auckland   Bangkok   Buenos Aires   Cape Town   Chennai
Dar es Salaam   Delhi   Hong Kong   Istanbul   Karachi   Kolkata
Kuala Lumpur   Madrid   Melbourne   Mexico City   Mumbai   Nairobi
São Paulo   Shanghai   Taipei   Tokyo   Toronto

Published by Oxford University Press, Inc.
198 Madison Avenue, New York, New York, 10016

www.oup.com

Oxford is a registered trademark of Oxford University Press

Library of Congress Cataloging-in-Publication Data
Cellular computing / edited by Martyn Amos.
p.   cm. — (Series in Systems Biology)
Includes bibliographical references and index.
ISBN 0-19-515539-4; 0-19-515540-8 (pbk)
1. Bioinformatics.   2. Cellular automata.   3. Molecular computers.   4. Nanotechnology.
I. Amos, Martyn.   II. Series.
QH324.2.C35 2004
571.6—dc22   2003058013

Series logo concept by Cloe L. Shasha.

9 8 7 6 5 4 3 2 1
Printed in the United States of America
on acid-free paper

Preface

The field of cellular computing is a novel and exciting development at the intersection of biology, computer science, mathematics, and engineering. Practitioners in this emerging discipline are concerned with the analysis, modeling, and engineering of inter- and intra-cellular processes for the purposes of computation. Developments in the field have potentially huge significance, ranging from new biological sensors and methods for interfacing living material with silicon substrates, through intelligent drug delivery and nanotechnology and on toward a deeper understanding of the fundamental mechanisms of life itself. This book provides both an introduction to some of the early fundamental work and a description of ongoing cutting-edge research in the field.

This volume is organized into three parts.

PART I: THEORETICAL AND ENGINEERING PRINCIPLES

Part I is concerned with the theoretical foundations of the field; in it we define the engineering principles upon which the second part is founded. Chapter 1 (Amos and Owenson) introduces the field of cellular computing, placing it in historical context and highlighting some of the more significant early papers in the field. A brief introduction to some underlying biological principles is supplied for the benefit of the nonspecialist biologist. Chapter 2 (Paton, Fisher, Malcolm, and Matsuno) reviews computational methods that may be useful to biologists seeking to model the interaction of proteins in spatially heterogenous and changing environments. Chapter 3 (Lones and Tyrell) describes enzyme genetic programming, a new optimization method that draws inspiration from biological representations of information. In Chapter 4 (Weiss, Knight, and Sussman) the idea of genetic process engineering is introduced. This is a methodology for mapping digital circuitry onto genetic elements in a rigorous and robust fashion. The fundamental engineering principles are established, and software to support development is described. Part I closes with Chapter 5 (Simpson et al.), in which whole cells are considered as analogous to semiconductor components. Communication between cells, and between cells and synthetic devices, is discussed, as well as their integration with nano- and micro-structured substrates and the modeling and simulation of relevant cellular processes.

PART II: LABORATORY EXPERIMENTS

Part II is concerned with reporting the results of laboratory experiments in cellular computing. In Chapter 6 (Wakabayashi and Yamamura), the construction of a bacterial logical inverter is described. This is developed further in Chapter 7, where Weiss, Knight, and Sussman describe the feasibility of digital computation in cells by building several *in vivo* digital logic circuits. Engineered intercellular communication, which may be crucial to the large-scale scalability of cellular computing, is also described. Chapter 8 (Simpson et al.) describes exciting work on the integration of living cells with micro- and nano-scale systems. This chapter includes state-of-the-art results concerning the incorporation of nanofibers into living cells.

PART III: COMPUTATION IN CILIATES

Part III covers an intriguing subfield of cellular computing concerned with the class of organisms known as ciliates. These organisms are particularly interesting because they "encrypt" their genomic information. Although studies of this phenomenon have not yet suggested obvious potential for human engineering of ciliates, elucidation of the underlying cellular processes will be of great use in the future. In Chapter 9, Prescott and Rozenberg both describe the assembly of ciliate genes from a biological perspective and consider its computational implications. The latter theme is developed further in Chapter 10, where Kari and Landweber study the decryption of ciliate DNA from a computational perspective.

ACKNOWLEDGMENTS

First thanks must go to the contributors. Their timeliness, flexibility, and helpfulness made editing this volume an enjoyable and rewarding experience. Thanks go also to Dennis Shasha, the series editor, and to Kirk Jensen at OUP for their advice and careful stewardship of the project. I am also grateful to Robert Heller for invaluable typesetting advice.

Martyn Amos
Exeter

Contents

Contributors

Martyn Amos
Department of Computer Science
School of Engineering, Computer
  Science and Mathematics
University of Exeter
UK

Chris D. Cox
Center for Environmental Biotechnology
University of Tennessee
USA

Michael Fisher
School of Biological Sciences
University of Liverpool
UK

James T. Fleming
Center for Environmental Biotechnology
University of Tennessee
USA

Michael A. Guillorn
Molecular-Scale Engineering and
  Nanoscale Technologies Research
  Group
Oak Ridge National Laboratory
USA

Lila Kari
Department of Computer Science
University of Western Ontario
Canada

Thomas F. Knight Jr.
Artificial Intelligence Laboratory and
  Department of Electrical Engineering
  and Computer Science
Massachusetts Institute of Technology
USA

Laura F. Landweber
Department of Ecology and Evolutionary
  Biology
Princeton University
USA

Michael A. Lones
Bio-Inspired Electronics Laboratory
Department of Electronics
University of York
UK

Grant Malcolm
Department of Computer Science
University of Liverpool
UK

Koichiro Matsuno
Department of BioEngineering
Nagaoka University of Technology
Japan

Timothy E. McKnight
Molecular-Scale Engineering and
    Nanoscale Technologies Research
    Group
Oak Ridge National Laboratory
USA

Anatoli Melechko
Center for Environmental Biotechnology
University of Tennessee
USA

Vladimir I. Merkulov
Molecular-Scale Engineering and
    Nanoscale Technologies Research
    Group
Oak Ridge National Laboratory
USA

Gerald Owenson
Biotechnology and Biological Sciences
    Research Council
(formerly Department of Biological
    Sciences, University of Warwick)
UK

Ray Paton
Department of Computer Science
University of Liverpool
UK

David M. Prescott
Department of Molecular, Cellular and
    Developmental Biology
University of Colorado
USA

Grzegorz Rozenberg
Department of Computer Science
University of Colorado
USA,
and
Leiden Institute of Advanced Computer
    Science
Leiden University
The Netherlands

John Sanseverino
Center for Environmental Biotechnology
University of Tennessee
USA

Gary S. Sayler
Center for Environmental Biotechnology
University of Tennessee
USA

Michael L. Simpson
Molecular-Scale Engineering and
    Nanoscale Technologies Research
    Group
Oak Ridge National Laboratory
USA

Gerald Sussman
Department of Electrical Engineering
    and Computer Science
Massachusetts Institute of Technology
USA

Andy M. Tyrell
Bio-Inspired Architectures Laboratory
Department of Electronics
University of York
UK

Kenichi Wakabayashi
Department of Computational
    Intelligence and Systems Science
Tokyo Institute of Technology
Japan

Ron Weiss
Department of Electrical Engineering
Princeton University
USA

Masayuki Yamamura
Department of Computational
    Intelligence and Systems Science
Tokyo Institute of Technology
Japan

Part I

Theoretical and Engineering Principles

1

An Introduction to Cellular Computing

Martyn Amos and Gerald Owenson

The abstract operation of complex natural processes is often expressed in terms of networks of computational components such as Boolean logic gates or artificial neurons. The interaction of biological molecules and the flow of information controlling the development and behavior of organisms is particularly amenable to this approach, and these models are well established in the biological community. However, only relatively recently have papers appeared proposing the use of such systems to perform useful, human-defined tasks. Rather than merely using the network analogy as a convenient technique for clarifying our understanding of complex systems, it is now possible to harness the power of such systems for the purposes of computation. The purpose of this volume is to discuss such work. In this introductory chapter we place this work in historical context and provide an introduction to some of the underlying molecular biology. We then introduce recent developments in the field of cellular computing.

INTRODUCTION

Despite the relatively recent emergence of molecular computing as a distinct research area, the link between biology and computer science is not a new one. Of course, for years biologists have used computers to store and analyze experimental data. Indeed, it is widely accepted that the huge advances of the Human Genome Project (as well as other genome projects) were only made possible by the powerful computational tools available to them. Bioinformatics has emerged as the science of the 21st century, requiring the contributions of truly interdisciplinary scientists who are equally at home at the lab bench or writing software at the computer.

However, the seeds of the relationship between biology and computer science were sown long ago, when the latter discipline did not even exist. When, in the 17th century, the French mathematician and philosopher René Descartes declared to Queen Christina of Sweden that animals could be considered a class of machines, she challenged him to demonstrate how a clock could reproduce. Three centuries later, with the publication of *The General and Logical Theory of Automata* [19] John von Neumann showed how a machine could indeed construct a copy of itself. Von Neumann believed that the behavior of natural organisms, although orders of magnitude more complex, was similar to that of the most intricate machines of the day. He believed that life was based on logic.

In 1970, the Nobel laureate Jacques Monod identified specific natural processes that could be viewed as behaving according to logical principles: "The logic of biological regulatory systems abides not by Hegelian laws but, like the workings of computers, by the propositional algebra of George Boole" [16, p. 76; see also 15].

The concept of molecular complexes forming computational components was first proposed by Richard Feynman in his famous talk "There's Plenty of Room at the Bottom" [11]. The idea was further developed by Bennett [6] and Conrad and Liberman [9], and since then there has been an explosion of interest in performing computations at a molecular level. In 1994, Adleman showed how a massively parallel random search may be implemented using standard operations on strands of DNA [1; see also 2]. Several authors have proposed simulations of Boolean circuits in DNA [3, 17], and recently the regulation of gene expression in bacteria has been proposed as a potential *in vivo* computational framework. We now discuss this last development in more detail.

BACKGROUND

Although proposed by Feynman [11] as long ago as 1959, the realization of performing computations at a molecular level has had to wait for the development of the necessary methods and materials. However, a rich body of theoretical work existed prior to Adleman's experiment. In 1982, Bennett [6] proposed the concept of a "Brownian computer" based around the principle of reactant molecules touching, reacting, and effecting state transitions due to their random Brownian motion. Bennett developed this idea by suggesting that a Brownian Turing machine could be built from a macromolecule such as RNA. "Hypothetical enzymes," one for each transition rule, catalyze reactions between the

RNA and chemicals in its environment, transforming the RNA into its logical successor.

Conrad and Liberman [9] developed this idea further, describing parallels between physical and computational processes (e.g., biochemical reactions being used to implement basic switching circuits). They introduced the concept of molecular level "word-processing" by describing it in terms of transcription and translation of DNA, RNA processing, and genetic regulation. However, their article lacks a detailed description of the biological mechanisms highlighted and their relationship with "traditional" computing. As the authors acknowledge, "Our aspiration is not to provide definitive answers . . . but rather to show that a number of seemingly disparate questions must be connected to each other in a fundamental way" [9, p. 240].

Conrad [8] expanded on this work, showing how the information processing capabilities of organic molecules may, in theory, be used in place of digital switching components (Figure 1.1a). Enzymes may cleave specific substrates by severing covalent bonds within the target molecule. For example, restriction endonucleases cleave strands of DNA at specific points known as *restriction sites*.

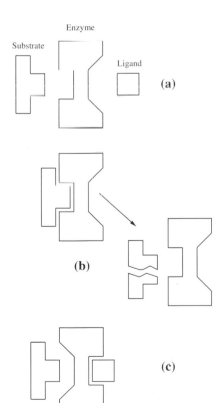

Figure 1.1 (a) Components of enzymatic switch. (b) Enzyme recognizes substrate and cleaves it. (c) Ligand binds to enzyme, changing its conformation. Enzyme no longer recognizes substrate.

In doing so, the enzyme switches the state of the substrate from one to another. Before this process can occur, a recognition process must take place, where the enzyme distinguishes the substrate from other, possibly similar molecules. This is achieved by virtue of what Conrad refers to as the "lock-key" mechanism, whereby the complementary structures of the enzyme and substrate fit together and the two molecules bind strongly (Figure 1.1b). This process may, in turn, be affected by the presence or absence of ligands. Allosteric enzymes can exist in more than one conformation (or "state"), depending on the presence or absence of a ligand. Therefore, in addition to the active site of an allosteric enzyme (the site where the substrate reaction takes place), there is a ligand binding site, which, when occupied, changes the conformation and hence the properties of the enzyme. This gives an additional degree of control over the switching behavior of the entire molecular complex.

In 1991, Hjelmfelt et al. [14] highlighted the computational capabilities of certain biochemical systems, as did Arkin and Ross in 1994 [4]. In 1995, Bray [7] discussed how the primary function of many proteins in the living cell appears to be the transfer and processing of information rather than metabolic processing or cellular construction. Bray observed that these proteins are linked into circuits that perform computational tasks such as amplification, integration, and intermediate storage.

GENETIC REGULATORY NETWORKS

In this section we develop the notion of circuits being encoded in genetic regulatory networks rather than simply being used as a useful metaphor. The central dogma of molecular biology is that DNA produces RNA, which in turn produces proteins. The basic building blocks of genetic information are known as *genes*. Each gene codes for a specific protein, and these genes may (simplistically) be turned on (*expressed*) or off (*repressed*). For the DNA sequence to be converted into a protein molecule, it must be read (*transcribed*) and the transcript converted (*translated*) into a protein (Figure 1.2).

Transcription of a gene produces a messenger RNA (mRNA) copy, which can then be translated into a protein. This results in the DNA containing the information for a vast range of proteins (*effector molecules*), but only those that are being expressed are present as mRNA copies. Each step of the conversion, from stored information (DNA) through mRNA (messenger) to protein synthesis (effector), is catalyzed by effector molecules. These effector molecules may be enzymes or other factors that are required for a process to continue. Consequently, a loop is formed, where products of one gene are required to produce further gene products, which may even influence that gene's own expression.

Genes are composed of a number of distinct regions that control and encode the desired product. These regions are generally of the form promoter–gene–

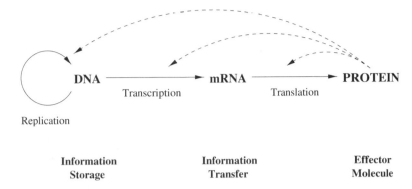

Figure 1.2 The central dogma of molecular biology.

terminator (Figure 1.3). Transcription may be regulated by effector molecules known as *activators* and *repressors*, which interact with the promoter and increase or decrease the level of transcription. This allows effective control over the expression of proteins, avoiding the production of unnecessary compounds.

An *operon* is a set of functionally related genes with a common promoter. An example of this is the *lac* operon, which contains three structural genes that allow *E. coli* to utilize the sugar lactose. When *E. coli* is grown on the common carbon source glucose, the product of the *lacI* gene represses the transcription of the *lacZYA* operon (Figure 1.4). However, if lactose is supplied together with glucose, a lactose by-product is produced that interacts with the repressor molecule, preventing it from repressing the *lacZYA* operon. This de-repression does not initiate transcription, since it would be inefficient to utilize lactose if the more common sugar glucose were still available. The operon is positively regulated by the CAP–cAMP (catabolite activator protein: cyclic adenosine monophosphate) complex, whose level increases as the amount of available glucose decreases. Therefore, if lactose were present as the sole carbon source, the *lacI* repression would be relaxed and the high CAP–cAMP levels would activate transcription, leading to the synthesis of the *lacZYA* gene products (Figure 1.5). Thus, the promoter is under the control of two sugars, and the *lacZYA* operon is only transcribed when lactose is present and glucose is absent. It is clear, therefore, that we may view the *lac* operon in terms of a genetic switch that is under the control of two sugars, lactose and glucose.

Figure 1.3 Major regions found within a bacterial operon.

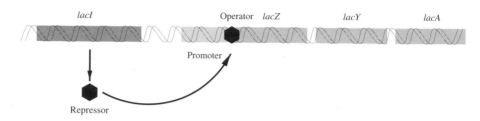

Figure 1.4 Repression of the *lac* operon by the *lacI* gene product.

ENCODING NETWORKS

So far, we have used computation only as a metaphor for intracellular processes. Since Adleman's original experiment, several other attempts to implement computations using DNA have been reported [5, 13, 20]. However, these experiments, although different in many ways, are characterized by the fact that they are all implemented *in vitro*: information is encoded as strands of DNA, and these are then manipulated in solution to perform a computation. It may be argued that this approach is suboptimal in that it fails to utilize the true potential of the DNA molecule in its natural environment (i.e., in the cell). The advantages of working *in vivo* as opposed to *in vitro* are numerous; rather than using DNA as a passive information carrier, we may take advantage of the fact that it can also be meaningful in a biological context. By reprogramming part of the cellular machinery to our advantage, the DNA "program" can affect its own execution.

In 1999, Weiss et al. [21] described a technique for mapping digital logic circuits onto genetic regulatory networks such that the resulting chemical activity within the cell corresponds to the computations specified by the digital

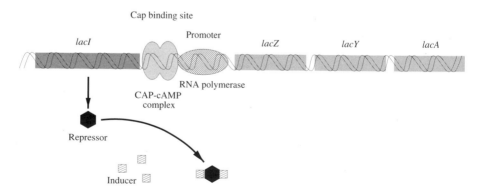

Figure 1.5 Positive control of the *lac* operon.

circuit (this work is described in chapter 4). There was a burst of activity in 2000, when two papers appeared in the same issue of *Nature*—both seminal contributions to the field. Elowitz and Leibler [10] described the construction of an oscillator network that periodically caused a culture of *E. coli* to glow by expressing a fluorescent protein. Crucially, the period of oscillation was slower than the cell division cycle, indicating that the state of the oscillator is transmitted from generation to generation. Gardner et al. [12] implemented a genetic toggle switch in *E. coli*. The switch is flipped from one stable state to another by either chemical or heat induction.

These single-cell experiments demonstrated the feasibility of implementing artificial logical operations using genetic modification. Savageau [18] addressed the issue of finding general design principles among microbial genetic circuits, citing several examples. This theme is developed in Part II of the current volume by establishing the field as a rigourous engineering discipline.

CONCLUSIONS

In this chapter we set the scene for what subsequently follows by providing an introduction to the emerging field of cellular computing and by summarizing some of some basic underlying biological principles. We traced the development of the field, placing it in historical context and highlighting landmark works. What was, until very recently, a field of theoretical interest has now become a realistic engineering discipline. As Michael Simpson acknowledges in chapter 5, this field is still at a very early stage in its development. The challenges facing researchers are significant but by no means insurmountable. If progress is to be made, it is clear that contributions will be needed from many disciplines, including the subdisciplines of biology as well as computer science, mathematics, engineering, and chemistry. This book represents one such multidisciplinary integration.

Acknowledgments   This work was partially supported by the European Commission IST network MolCoNet, contract number IST-2001-32008. Gerald Owenson was previously employed by the University of Warwick, UK.

References

[1] Leonard M. Adleman. Molecular computation of solutions to combinatorial problems. *Science*, 266:1021–1024, 1994.

[2] Martyn Amos. DNA computation. PhD thesis, Department of Computer Science, University of Warwick, UK, 1997.

[3] Martyn Amos, Paul E. Dunne, and Alan Gibbons. DNA simulation of Boolean circuits. In John R. Koza, Wolfgang Banzhaf, Kumar Chellapilla, Kalyanmoym

Deb, Marco Dorigo, David B. Fogel, Max H. Garzon, David E. Goldberg, Hitoshi Iba, and Rick Riolo, editors, *Genetic Programming 1998: Proceedings of the Third Annual Conference*, pages 679–683. Morgan Kaufmann, San Francisco, CA, 1998.

[4] A. Arkin and J. Ross. Computational functions in biochemical reaction networks. *Biophys. J.*, 67:560–578, 1994.

[5] Yaakov Benenson, Tamar Paz-Elizur, Rivka Adar, Ehud Keinan, Zvi Livneh, and Ehud Shapiro. Programmable and autonomous computing machine made of bio-molecules. *Nature*, 414:430–434, 2001.

[6] C. H. Bennett. The thermodynamics of computation—a review. *Intl. J. Theor. Phys.*, 21:905–940, 1982.

[7] Dennis Bray. Protein molecules as computational elements in living cells. *Nature*, 376:307–312, 1995.

[8] Michael Conrad. On design principles for a molecular computer. *Commun. ACM*, 28:464–480, 1985.

[9] Michael Conrad and E. A. Liberman. Molecular computing as a link between biological and physical theory. *J. Theor. Biol.*, 98:239–252, 1982.

[10] M. B. Elowitz and S. Leibler. A synthetic oscillatory network of transcriptional regulators. *Nature*, 403:335–338, 2000.

[11] Richard P. Feynman. There's plenty of room at the bottom. In D. Gilbert, editor, *Miniaturization*, pages 282–296. Reinhold, New York, 1961.

[12] T. S. Gardner, C. R. Cantor, and J. J. Collins. Construction of a genetic toggle switch in *Escherichi coli*. *Nature*, 403:339–342, 2000.

[13] Frank Guarnieri, Makiko Fliss, and Carter Bancroft. Making DNA add. *Science*, 273:220–223, 1996.

[14] Allen Hjelmfelt, Edward D. Weinberger, and John Ross. Chemical implementation of neural networks and Turing machines. *Proc. Natl. Acad. Sci. USA*, 88:10983–10987, 1991.

[15] F. Jacob and J. Monod. Genetic regulatory mechanisms in the synthesis of proteins. *J. Mol. Biol.*, 3:318–356, 1961.

[16] Jacques Monod. *Chance and Necessity*. Penguin, London, 1970.

[17] Mitsunori Ogihara and Animesh Ray. Simulating Boolean circuits on a DNA computer. In *Proceedings of the First Annual International Conference on Computational Molecular Biology (RECOMB97)*, pages 226–231. ACM Press, New York, 1997.

[18] Michael A. Savageau. Design principles for elementary gene circuits: Elements, methods and examples. *Chaos*, 11(1):142–159, 2001.

[19] John von Neumann. The general and logical theory of automata. In L. A. Jeffress, editor, *Cerebral Mechanisms in Behavior*, pages 1–41. Wiley, New York, 1941.

[20] Liman Wang, Jeff G. Hall, Manchun Lu, Qinghua Liu, and Lloyd M. Smith. A DNA computing readout operation based on structure-specific cleavage. *Nature Biotechnol.*, 19:1053–1059, 2001.

[21] R. Weiss, G. Homsy, and T. F. Knight Jr. Toward *in-vivo* digital circuits. In L. F. Landweber and E. Winfree, editors, *DIMACS Workshop on Evolution as Computation*, pages 275–297. Springer-Verlag, New York, 1999.

2

Proteins and Information Processing

Ray Paton, Michael Fisher, Grant Malcolm,
and Koichiro Matsuno

This chapter reviews and briefly discusses a set of computational methods that can assist biologists when seeking to model interactions between components in spatially heterogeneous and changing environments. The approach can be applied to many scales of biological organization, and the illustrations we have selected apply to networks of interaction among proteins.

INTRODUCTION

Biological populations, whether ecological or molecular, homogeneous or heterogeneous, moving or stationary, can be modeled at different scales of organization. Some models can be constructed that focus on factors or patterns that characterize the population as a whole such as population size, average mass or length, and so forth. Other models focus on values associated with individuals such as age, energy reserve, and spatial association with other individuals. A distinction can be made between population (p-state) and individual (i-state) variables and models. We seek to develop a general approach to modeling biosystems based on individuals.

Individual-based models (IBMs) typically consist of an environment or framework in which interactions occur and a number of individuals defined in terms of their behaviors (such as procedural rules) and characteristic parameters. The actions of each individual can be tracked through time. IBMs represent heterogeneous systems as sets of nonidentical, discrete, interacting, autonomous, adaptive agents (e.g., Devine and Paton [5]). They have been used to model the dynamics of population interaction over time in ecological systems, but IBMs can equally be applied to biological systems at other levels of scale. The IBM approach can be used to simulate the emergence of global information processing from individual, local interactions in a population of agents.

When it is sensible and appropriate, we seek to incorporate an ecological and social view of inter-agent interactions to all scales of the biological hierarchy [6, 12, 13]. In this case we distinguish among individual "devices" (agents), networks (societies or communities), and networks in habitats (ecologies). In that they are able to interact with other molecules in subtle and varied ways, we may say that many proteins have social abilities [22, 25]. This social dimension to protein agency also presupposes that proteins have an underlying ecology in that they interact with other molecules including substrates, products, regulators, cytoskeleton, membranes, water, and local electric fields. The metaphor also facilitates a richer understanding of the information-processing capacities of cells [11, 12, 24]. This can be characterized as an ability to act in a flexible, unscripted manner—another feature of adaptive agents.

SCOPE FOR MICROPHYSICAL INFORMATION PROCESSING IN PROTEINS

Following an established tradition that can be traced back to Rosen [17], we support the view that information is generated at the quantum level and manifested at mesoscopic and macroscopic levels within molecular and cellular systems. Quantum effects can be related to many biological processes. Clearly, interactions between photons and matter are quantum-mechanical in nature, and so we may think about "biophotons," bioluminescence, photosynthesis, and photodetection. When molecular interactions occur in proteins and polynucleotides, quantum processes are taking place; these can be related to shape-based interactions and molecular recognition as well as to more long-range phenomena. Cellular microenvironments are very far removed from *in vitro* homogeneous high-dilution experimental systems. They are highly structured, with (relatively) low local water content and complex microarchitectures. A number of molecules and molecular systems that could form part of cellular quantum information-processing systems may be described. Components for biological quantum information processing could include wiring (e.g., conductive biopolymers); storage (e.g., photosystem II reaction centers, cytochromes, blue proteins, ferritin); and gates and switches such as bacteriorhodopsin, cell receptors, and ATPase. Solitonlike mechanisms may result in the conduction of electrons and bond vibrations along sections of alpha-helices, and Ciblis and

Cosic [1] discussed the potential for vibrational signaling in proteins. Conrad [3] described the quantum level by elaborating the notion of vertical information processing in biological systems.

Welch [23] considered an analogue field model of the metabolic state of a cell based on ideas from quantum field theory. He proposed that the structure of intracellular membranes and filaments, which are fractal in form, might generate or sustain local fields. Virtually all biomembranous structures *in vivo* can generate local electric fields and proton gradients. Enzymes can act as the energy-transducing measuring devices of such local fields. In some ways we may say that the field provides a "glue," which was not available at the individual, localized level of discrete components (see also Paton [13]). Popp et al. [16] discussed the possibility of DNA acting as a source of lased "biophotons." This was based on experiments in which DNA conformational changes induced with ethidium bromide *in vivo* were reflected by changes in the photon emission of cells. In another study, Popp et al. [15] compared theoretically expected results of photon emission from a chaotic (thermal) field with those of an ordered (fully coherent) field with experimental data and concluded that there are ample indications for the hypothesis that biophotons originate from a coherent field within living tissues.

Klinman [10] discussed *in vitro* experiments of hydrogen tunneling at room temperature in yeast alcohol dehydrogenase and bovine serum amine oxidase. She showed that the reaction coordinate for these enzymes, rather than being a sharp transition (giving a step function) is smoothed to give a sigmoidal/logistic-shaped curve. These molecules are measuring quantum effects that are magnified to the meso- and macroscale. Given that enzyme–substrate complexes and many other protein-based interactions provide switching functions, we here have an example of a quantum mechanical switch, albeit within a test tube rather than an intracellular experiment. These enzymes are fuzzy not just because of thermodynamic effects but because of interactions and measurements taking place at the microscale. This capacity for interaction implies local measurement and information generation.

PROTEINS AS "SMART" COMPUTATIONAL AGENTS

> A goal for the future would be to determine the extent of knowledge the cell has of itself and how it utilises this knowledge in a "thoughtful" manner when challenged.
>
> —Barbara McClintock

We noted in the previous section that biological information can be generated at the quantum (micro-) level of biological organization and that the effects impact on meso- and macroscales. Knowledge about the subtle intricacies of molecular processes in cells and their subcompartments is increasing rapidly. Cells are highly structured, hierarchically organized, open systems. We argue that contemporary models must take account of spatial heterogeneity, multifunctionality, and individuality.

Conrad [4] discussed the idea of a seed-germination model of enzyme action. This model sought to take account of the multiplicities of interaction that give rise to enzyme function—the factor complex associated with action. Coupled with this interactional view is the self-assembly paradigm [3], which also addresses issues of nonprogrammability [2]. Enzymes and other proteins can be described as a "smart thermodynamic machines" which satisfy a "gluing" (functorial) role in the information economy of the cell [14]. We exploit these views by drawing comparisons between enzymes and verbs. This text/dialogical metaphor also helps refine our view of proteins as context-sensitive information processing agents [13].

Many proteins display a number of "cognitive" capacities, including

- pattern recognition
- multifunctionality
- handling fuzzy data
- memory capacity
- signal amplification
- integration and crosstalk
- context-sensitivity.

Multifunctionality is also a dimension of cognitive capacity. Transcription factors such as CBP and p300 have multiple functions associated with them, including molecule–molecule binding and interactions, enzymatic processes, physical bridges between various transcription factors and the basal transcriptional machinery, acting as histone acetyltransferases (HATs)—linking transcription to chromatin remodeling, and mediating negative and positive crosstalk between different signaling pathways.

The information-processing nature of eukaryotic intracellular signaling pathways illustrates many of these concepts well. In the classical model, a cell-surface receptor binds an extracellular effector (e.g., hormone, pheromone). Receptor occupation is transduced into an intracellular signal by activation of an effector enzyme (e.g., adenylate cyclase, phospholipase) responsible for synthesis of a secondary messenger (e.g., cyclic AMP, diacylglycerol). The secondary messenger then promotes activation/inactivation of protein kinases and/or protein phosphatases. Subsequent changes in the phosphorylation state of target phosphoproteins (e.g., enzymes, structural elements, transcription factors) bring about the changes in cellular activity observed in response to the external signal (see Figure 2.1).

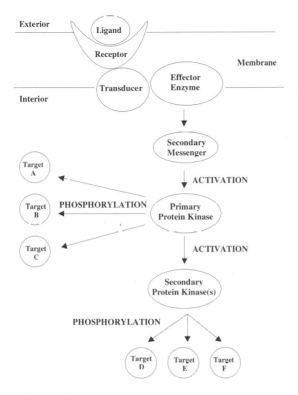

Figure 2.1 Classic second messenger system.

Intracellular signaling pathways all share important information-processing features; for example, the generation of a secondary messenger and occurrence of cascades, or hierarchies, of protein kinases permits a considerable degree of amplification to be introduced. The classical model for intracellular signaling therefore presents a highly sensitive mechanism for relaying small changes in the external environment to the interior of the cell. The system is highly flexible and is easily adapted to respond to a diverse range of primary messenger/receptor interactions. A key feature of many signaling pathways, not initially apparent in this model, is the ability to deliver receptor-derived information to unique sub-sets of target phosphoproteins. For example, although many different hormones and effectors use a common cyclic AMP-based pathway to activate the cyclic AMP-dependent protein kinase (PKA), the consequences of PKA activation can be very different. Specificity is built into the signaling pathway through the activation of spatially distinct subsets of PKA molecules. Spatial organization of PKA is therefore a major regulatory mechanism for ensuring the selectivity and specificity of cyclic AMP-mediated responses.

As indicated above, it is possible to consider the protein components of signaling pathways to be "smart thermodynamic machines" [7]. In this respect, protein kinases and protein phosphatases display "cognitive" capacities such as pattern recognition, ability to handle fuzzy data, memory capacity, and context sensitivity [7]. For example, the major signal-sensitive protein kinases (PKA, PKC, and calmodulin-dependent kinase II [CaMK]) are obviously all catalysts of phosphorylation. Additionally, they are all switches that can be activated by the appropriate secondary messenger (cyclic AMP/PKA; diacylglycerol/PKC; Ca^{2+}/CaMK). Specific isoforms of these enzymes may also be subject to autophosphorylation. Phosphorylation of the RII isoform of the PKA regulatory subunit prolongs the dissociated, activated state of PKA [21]. Similarly, CaMK carries out autophosphorylation of an inhibitory domain, thereby prolonging the activated state of the enzyme [19]. As a consequence, protein kinases can be considered to have a capacity for memory (i.e., even when secondary messenger signals have diminished, their phosphorylating power is preserved). Protein kinases and protein phosphatases may also possess positional or targeting information [9]. For example, isoforms of the PKA catalytic subunit can be modified by the addition of a myristoyl group. This fatty acid-derived group may direct PKA to specific membrane locations, or alternatively, into specific protein–protein interactions. The spatial organization of PKA also depends on its association with structural elements of the cell via anchor proteins, or AKAPs—A kinase anchor proteins [18]. Specific isoforms of CaMK also possess positional information in that nuclear-specific localization sequences target this enzyme to the cell nucleus, and, consequently, CaMKs play a role in the phosphorylation of proteins involved in the control of gene expression [8].

Perhaps the most sophisticated example of spatial organization of signaling pathway components concerns the mitogen-activated protein kinase cascades (MAPK cascades). The spatial organization of these protein kinase cascades leads to distinct cellular responses to a diverse group of environmental stimuli [20]. MAPK cascades are organized into discrete, parallel signaling complexes by scaffold proteins; the signaling cascade components are brought into close physical contact allowing rapid and direct transfer of signaling information (see Figure 2.2). An intriguing feature of these signaling pathways is that, despite sharing common components, they are normally extremely well insulated from each other and show little if any crosstalk or cross-activation [6].

CONCLUSIONS

We have given an overview of our approach to individual-based modeling of interactions between components in spatially heterogeneous and changing environments. The key perspective in our discussion has been on protein interactions as manifesting information-processing capabilities. The notion of

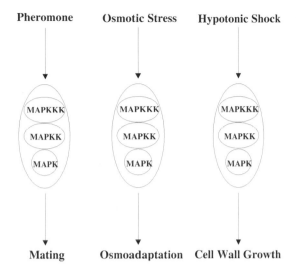

Figure 2.2 Parallel MAP kinase cascades in yeast.

information ranges from the quantum-mechanical to the social or ecological, and we have illustrated our view of proteins as smart computational agents with a discussion of cascades in intracellular signaling pathways. Further aspects of the mathematical and computational modeling of these processes are described by Fisher et al. [6].

References

[1] P. Ciblis and I. Cosic. The possibility of soliton/exciton transfer in proteins. *J. Theor. Biol.*, 184:331–338, 1997.

[2] M. Conrad. The price of programmability. In H. Herken, editor, *The Universal Turing Machine: A Fifty Year Survey*, pages 285–307. Oxford University Press, Oxford, 1988.

[3] M. Conrad. Quantum mechanics and cellular information processing: the self-assembly paradigm. *Biomed. Biochim. Acta*, 49(8-9):743–755, 1990.

[4] M. Conrad. The seed germination model of enzyme catalysis. *BioSystems*, 27:223–233, 1992.

[5] P. Devine and R. C. Paton. Biologically-inspired computational ecologies: a case study. In David Corne and Jonathan L. Shapiro, editors, *Evolutionary Computing, AISB Workshop*, vol. 1305, *Lecture Notes in Computer Science*, pages 11–30. Springer, Berlin, 1997.

[6] M. Fisher, G. Malcolm, and R. C. Paton. Spatio logical processes in intracellular signalling. *BioSystems*, 55:83–92, 2000.

[7] M. Fisher, R. C. Paton, and K. Matsuno. Intracellular signalling proteins as 'smart' agents in parallel distributed processes. *BioSystems*, 50:159–171, 1999.

[8] E. K. Heist and H. Schulman. The role of Ca^{2+}/calmodulin-dependent proteins within the nucleus. *Cell Calcium*, 23:103–114, 1998.

[9] M. Hubbard and P. Cohen. On target with a mechanism for reversible phosphorylation. *Trends Biochem. Sci.*, 18:172–177, 1993.

[10] J. P. Klinman. Quantum mechanical effects in enzyme-catalysed hydrogen transfer reactions. *Trends Biochem. Sci*, 14:368–373, 1989.

[11] P. C. Marijuan. 'gloom in the society of enzymes': on the nature of biological information. *BioSystems*, 38(2/3):163–171, 1996.

[12] R. C. Paton. Some computational models at the cellular level. *BioSystems*, 29:63–75, 1993.

[13] R. C. Paton. Glue, verb and text metaphors in biology. *Acta Biotheor.*, 45:1–15, 1997.

[14] R. C. Paton and K. Matsuno. Some common themes for enzymes and verbs. *Acta Biotheor.*, 46:131–140, 1998.

[15] F. A. Popp, K. H. Li, W. P. Mei, M. Galle, and R. Neurohr. Physical aspects of biophotons. *Experientia*, 44:576–585, 1988.

[16] F. A. Popp, W. Nagl, K. H. Li, W. Scholz, O. Weingarter, and R. Wolf. Biophoton emission. New evidence for coherence and DNA as source. *Cell. Biophys.*, 6:33–52, 1984.

[17] R. Rosen. A quantum-theoretic approach to genetic problems. *Bull. Math. Biophys.*, 22:227–255, 1960.

[18] C. S. Rubin. A kinase anchor proteins and the intracellular tergeting of signals carried by camp. *Biochim. Biophys. Acta*, 1224:467–479, 1994.

[19] H. Schulman and L. L. Lou. Multifunctional ca2+/calmodulin-dependent protein kinase: domain structure and regulation. *Trends Biochem. Sci.*, 14:62–66, 1989.

[20] G. F. Sprague. Control of MAP kinase signalling specificity or how not to go HOG wild. *Genes Devel.*, 12:2817–2820, 1998.

[21] K. Takio, S. B. Smith, E. G. Krebs, K. A. Walsh, and K. Titani. Amino acid sequence of the regulatory subunit of bovine type II adenosine cyclic $3',5'$ phosphate dependent protein kinase. *Biochemistry*, 23:4200–4206, 1984.

[22] G. R. Welch. The living cell as an ecosystem: hierarchical analogy and symmetry. *Trends Ecol. Evol.*, 2:255–267, 1987.

[23] G. R. Welch. An analogical 'field' construct in cellular biophysics: History and present status. *Prog. Biophys. Mol. Biol.*, 57:71–128, 1992.

[24] G. R Welch. The enzymatic basis of information processing in the living cell. *BioSystems*, 38(2/3):147–153, 1996.

[25] G. R. Welch and T. Keleti. Is cell metabolism controlled by a 'molecular democracy' or by a 'supramolecular socialism'? *Trends Biochem. Sci.*, 12:216–217, 1978.

3

Enzyme Genetic Programming

Michael A. Lones and Andy M. Tyrrell

Programming is a process of optimization; taking a specification, which tells us what we want, and transforming it into an implementation, a program, which causes the target system to do exactly what we want. Conventionally, this optimization is achieved through manual design. However, manual design can be slow and error-prone, and recently there has been increasing interest in automatic programming; using computers to semiautomate the process of refining a specification into an implementation.

Genetic programming is a developing approach to automatic programming, which, rather than treating programming as a design process, treats it as a search process. However, the space of possible programs is infinite, and finding the right program requires a powerful search process. Fortunately for us, we are surrounded by a monotonous search process capable of producing viable systems of great complexity: evolution.

Evolution is the inspiration behind genetic programming. Genetic programming copies the process and genetic operators of biological evolution but does not take any inspiration from the biological representations to which they are applied. It can be argued that the program representation that genetic programming does use is not well suited to evolution. Biological representations, by comparison, are a product of evolution and, a fact to which this book is testament, describe computational structures.

This chapter is about enzyme genetic programming, a form of genetic programming that mimics biological representations in an attempt to improve the evolvability of programs. Although it would be an advantage to have a familiarity with both genetic programming and biological representations, concise introductions to both these subjects are provided.

INTRODUCTION

Artificial Evolution and Evolutionary Computation

According to modern biological understanding, evolution is solely responsible for the complexity we see in the structure and behavior of biological organisms. Nevertheless, evolution itself is a simple process that can occur in any population of imperfectly replicating entities where the right to replicate is determined by a process of selection.

Consequently, given an appropriate model of such an environment, evolution can also occur within computers. This artificial evolution is the domain of a discipline called *evolutionary computation* (EC), which uses the search and optimization qualities of evolution to design, search for, and optimize any structure that can be represented on a computer. Programs that implement artificial evolution are called *evolutionary algorithms* (EAs). There were three original forms of EA: genetic algorithms [13, 14], evolution strategies [5, 33, 36], and evolutionary programming [12]. Although contemporary algorithms now form a continuum of approaches, these three terms are still commonly used as a basis for classification.

The activity of an EA revolves around a fundamental data structure called the *population*—a collection of individuals, each of which encodes a candidate solution to a particular problem. It is common to use biological terminology in EC. The value of a particular candidate solution, and a measure of how well it solves the problem, is called fitness. The encoded form of a solution is known as a genotype and the actual solution a phenotype.

The goal of an EA is to find an optimal solution to a problem. It does this through the iterative replication of existing, fit genetic material to form new, hopefully fitter, individuals that then replace less fit individuals within the population. Good genetic material is promoted by a selection process that determines which individuals contribute toward the genetic makeup of new individuals. To prevent genetic stagnation, new genetic material is introduced by making occasional random changes to the genotypes of new individuals, a process called mutation.

In some EAs mutation is the only genetic operator and new individuals are simply mutated versions of existing individuals. Other EAs, including genetic algorithms (GAs), use a crossover operator that generates new individuals by splicing together components from more than one existing solution. This resembles biological recombination and is theorized to encourage the formation and replication of high-fitness genetic components called building blocks.

Representation in Evolutionary Algorithms

The way in which a candidate solution is recorded is called its representation. For a particular problem domain there may be many ways of recording a solution. Choosing an appropriate representation is akin to the decision made by a programmer when deciding how to record a certain piece of information in memory. There may be a selection of representations available, such as arrays, linked lists, hash tables, and trees, and each of these will have different advantages and disadvantages depending on how the information will be accessed and used. The programmer must trade off the various pros and cons and select the one that is most effective or efficient. On many occasions this will not be the most natural or understandable representation.

In an EA, a representation is accessed and used in two different ways. During evaluation, the effectiveness of the solution is measured to gauge its fitness. During evolution, the solution is modified by variational operators such as mutation and crossover. This introduces two sets of demands upon a representation: generative and variational [2]. Generative demands depend on the domain, but in general it is required that the representation makes solutions easy to measure. Variational demands require that the representation is evolvable.

Evolvability is the ability of a representation, when subject to variational operators, to generate solutions of increasing fitness on a regular basis. The role of artificial evolution is to search within a space of possible solutions and find a solution that is optimal with respect to the fitness-measuring function. The manner, and hence the effectiveness, with which this space is searched is decided by the action of the variational operators upon current solutions within the population. The way in which the variational operators transform existing solutions determines the form and fitness of future solutions. Moreover, the way in which the variational operators transform existing solutions is determined by the degrees of freedom presented by the representation. Consequently, the representation determines which paths are available through the search space. An evolvable representation promotes paths that lead to better solutions and ultimately to optimal solutions.

Unfortunately, it is not easy to predetermine what constitutes a fitness-increasing transformation during evolution. It is, however, possible to identify what constitutes a bad transformation. Consequently, the incidence of fitness-increasing transformations can be increased by reducing the incidence of fitness-decreasing transformations.

The easiest way to enact a bad transformation is to make a large change to an existing solution. This is because a search space usually contains many more bad solutions than it does good solutions, and making a large change to a fitter-than-average solution will most likely move it to a lower fitness location in the search space. In fact, most changes to a fitter-than-average solution will lead to a less fit solution. However, the chance of finding a higher fitness solution can be increased by only making small changes to existing solutions.

In practical terms, this requires that operators generate new solutions (child solutions) that are similar to the existing solutions (parent solutions) they are derived from. For mutation, a small change to the information held within a representation should, in general, lead to a small change in the solution it represents. For crossover, which derives new solutions from more than one existing solution, the issues are more complicated. However, there is still a requirement for each child solution to bear reasonable similarity to one of its parents. A further requirement is that the recombination be meaningful, or to use a simple metaphor, a child should not consist of two heads and no body.

Genetic Programming

Artificial evolution can be applied to any structure that can be represented within a computer. This includes computer programs. Genetic programming (GP) [6, 15, 16, 17] is an approach in evolutionary computation that evolves programs and other executable structures. The original GP algorithm was designed by Koza [16], and many aspects of it, including the representation, remain standard in modern GP.

In Koza's GP, programs are represented by their parse tree. A parse tree captures the execution ordering of a program and shows the hierarchical relationships between components of the program. Functions occur at internal tree nodes and terminals occur at leaf nodes. For a hierarchical language such as LISP, the parse tree is the program. However, GP typically uses simple un-typed languages that can be fully described by a function set and a terminal set. Since there are no types, any function can receive the output of any other function or terminal as input. Consequently, any parse tree will be syntactically correct.

The initial population is filled with random parse trees of various shapes and sizes, each consisting of randomly chosen functions and terminals. The standard variational operators are mutation and crossover. There are two kinds of mutation. Point mutation replaces a single terminal with a single terminal or a single nonterminal with a single nonterminal. Subtree mutation replaces a subtree with a new randomly generated subtree. Crossover, which is depicted in Figure 3.1, creates child parse trees by swapping a pair of subtrees between two parent parse trees.

The Representation Problem of GP

Recall that evolvability places certain demands on recombination. First, new solutions should be similar to at least one of the solutions they were derived from. Second, recombination should be meaningful. In the example recombination shown in Figure 3.1, neither of these demands has been met, despite the fact that the parent solutions are quite similar to one another.

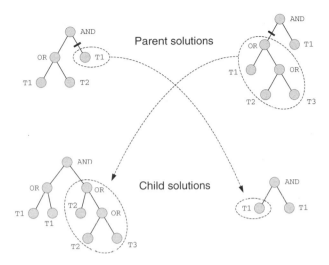

Figure 3.1 Subtree crossover. Crossover points are chosen randomly in two parent solutions, and the subtrees below the crossover points are swapped to produce child solutions.

Subtree crossover is a natural recombination operator for parse trees and can be useful to evolution. However, subtree crossover does have significant limitations [3, 31]. In one sense, the behaviors that subtree crossover can carry out are too constrained. This is because subtree crossover may only work with full subtrees. A group of components that do not form a complete subtree can only be exchanged if the entire subtree within which they reside is exchanged. This means that an incomplete subtree that does not have a fully ground set of leaf nodes may not be transferred, although this operation would be particularly useful for incomplete subtrees near the root of the parse tree. In this situation swapping the entire subtree with another would almost inevitably make the program dysfunctional. Figure 3.2 shows what an incomplete subtree crossover could look like.

In another sense, the behaviors that subtree crossover does carry out are too unconstrained. This is because subtrees are exchanged nondeterministically. When a functional subtree is replaced by a subtree that does not carry out a similar function, this will most likely cause the program to stop working. Hence, most nondeterministic subtree crossovers will produce inviable recombinants. Relatively few crossovers will have a positive, or at least neutral, effect [30]. Consequently, most of the behaviors carried out by subtree crossover are not useful most of the time.

One solution to these problems is to modify crossover to allow a greater range of behaviors while disallowing disruptive behaviors [20]. Such an operator

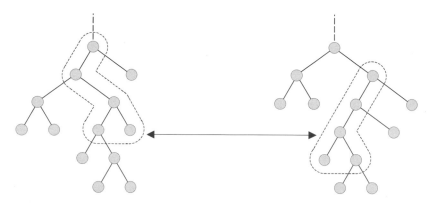

Figure 3.2 Subcomponent crossover within a parse tree. This kind of exchange is not possible with standard subtree crossover.

would have a considerable computational overhead. Moreover, the real problem does not lie with crossover but with the parse tree representation.

One problem is that the parse tree representation is not very flexible and does not readily support many forms of editing. Other, more flexible, representations have been used in GP [4, 26, 32, 35], though not widely. However, inflexibility is just part of a wider problem of the parse tree representation not being well suited to evolution. Parse trees were not designed to be evolvable, and therefore it is unsurprising that they are not ideal candidates for evolution. Evolvability is not a common property of representations, and it seems likely that it is especially uncommon among those used by humans to solve problems. Trees, for example, implicitly capture the needs of humans that a structure be decomposed and explicitly connected in order to be easily understood. Evolution does not require these constraints because it does not need to understand what is being evolved. Introducing constraints to evolution only serves to limit what can be evolved.

A Biomimetic Solution

The obvious solution, therefore, is to find a program representation that is evolvable. This is the approach taken by enzyme genetic programming. However, enzyme GP does not attempt to construct a new representation afresh, but rather look to biology, an evolutionary system that already possesses an evolvable representation, which, in a sense, represents computational structures. The aim of enzyme GP is to identify principles of biological representations thought to contribute to their evolvability and adapt these principles to improve the evolvability of representations in genetic programming. It should be stated that the aim is not to produce an exact model of these biological representations but only to mimic those constructs and behaviors that might improve GP. Before introducing enzyme GP, we review the biological representations on which it is based.

BIOLOGICAL REPRESENTATIONS

An organism's genome, the sum of all the DNA from all its chromosomes, encodes the organism's proteome, the set of all protein species capable of being created by the organism through the processes of DNA transcription and translation. Each protein is specified by a gene, a sequence of codons stating the order and nature of the components from which the protein is to be assembled, concluded by a termination codon, marking the end of the gene. Proteins have four layers of structure. Unfolded, they form a string of amino acids. Amino acids are chemicals unified by a common structure and differentiated by the chemical properties of their side chains. However, proteins do not remain unfolded. Secondary structure, the emergence of physical members and chemical surfaces, is caused by chemical attractions between the amino acids within the chain. This effect is compounded by the aqueous environment of the cell which, through interactions between the protein and water molecules, forces the structure into a distinct folded form known as the tertiary structure. The final layer of structure, quaternary, emerges from single-chain molecules bonding together to form macromolecular complexes.

The overall behavior of an organism does not emerge from isolated properties of individual proteins, but rather from the interactions between proteins and between proteins and other chemicals. These interactions lead to the formation of structures called biochemical pathways; sequences of protein-mediated reactions and interactions that carry out functional tasks within the organism. There are three broad categories of biochemical pathway, illustrated conceptually in Figure 3.3. Metabolic pathways exist within cells and emerge from the interactions of locally transcribed proteins. Signaling pathways comprise cellular responses to intercellular signals. Gene expression networks describe regulatory

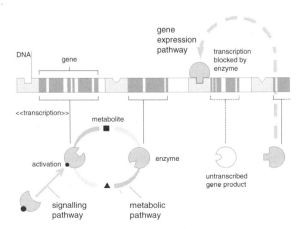

Figure 3.3 Biochemical pathways.

interactions between genes and gene products. Marijuán [27] describes these activities respectively, as *self-organizing*, *self-reshaping*, and *self-modification*. Self-organization is the ability of a distributed system to carry out a unified task. Self-reshaping is the ability of an existing system to carry out multiple tasks to satisfy varying needs. Self-modification is the ability of a system to change its constitution in order to solve unforseen problems. Interactions between the three classes of pathway unify these activities and bring out the emergent behavior of the whole.

Metabolic Pathways

Cellular computation takes the form of the metabolism of the cell's chemical state. Metabolism of this state involves transformations, implemented by constructive and destructive chemical reactions, between chemical species, increasing the concentrations of some and reducing the concentrations of others. Temperatures in biological systems are relatively low, and many of these reactions will not occur without the help of enzymes, catalytic proteins that bind to specific chemicals and mediate their interaction and subsequent transformation. The chemical species to which an enzyme binds, its substrates, are determined by a property called its specificity; a result of the spatial arrangement of amino acids found at the enzyme's binding sites. Substrate binding occurs through inexact matching. The closer the match is between the substrate shape and the binding site shape, the stronger the bond between enzyme and substrate and, consequently, the higher the likelihood of recognition.

The presence of enzymes activates transformative paths within the metabolism. This forms a network, a metabolic network, where chemical species are nodes and enzyme-mediated reactions are the connections between nodes. Metabolic networks are composed of metabolic pathways. A metabolic pathway is an assemblage of enzymes which, roughly speaking, carries out a unified task involving a shared group of chemical species. This cooperation emerges from the sharing of substrates between enzymes, where the product of one enzyme becomes the substrate of another.

Proteins, including enzymes, have computational characteristics similar to artificial computational elements like transistors and logic gates [8, 9, 11] (and some that are quite unrelated [10]). The computation provided by metabolic pathways has been modeled using Petri nets [34], which are normally used to describe distributed computation in artificial systems. In other work, artificial analogues of enzymes have been applied to information processing [37].

FUNDAMENTALS OF ENZYME GP

Enzyme GP [22–25] is a genetic programming system that evolves executional structures using a genetic algorithm. The biomimetic program representations that enzyme GP uses during evolution differ considerably from the parse tree structures used by conventional GP, yet enzyme GP still generates conventional program expressions during its evaluation phase.

A metabolic network is a group of enzymes that interact through product–substrate sharing to achieve some form of computation. In a sense, a metabolic network is a program where the enzymes are the functional units, and the interactions between enzymes are the flow of execution. Consequently, the set of possible interactions between enzymes determines the possible structures of the program. Enzyme GP represents programs as structures akin to metabolic networks. These structures are composed of simple computational elements modeled on enzymes. The elements available to a program are recorded as analogues of genes in an artificial genome. These genomes are then evolved with the aim of optimizing both the function and connectivity of the elements they contain.

From a nonbiological perspective, enzyme GP represents a program as a collection of components where each component carries out a function and interacts with other components according to its own locally defined interaction preferences. A program component is a terminal or function instance wrapped in an interface that determines both how the component appears to other components and, if the component requires input, which components it would like to receive input from.

Modeling the Enzyme

Enzyme GP captures the enzyme as a computational element that transforms the products of other enzymes and does so according to its predefined affinity for their products. The components of the model are depicted in Figure 3.4.

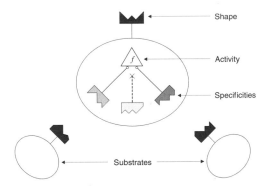

Figure 3.4 Enzyme model. An enzyme consists of an activity and a set of specificities. Shape describes how the enzyme is seen by other program components. Specificity determines, in terms of their shape, which components will be bound as substrates to provide input.

This artificial enzyme consists of an activity, a set of specificities, and a shape. Of these, activity and specificity are defined in the enzyme's gene. Shape is calculated from activity and specificity when the enzyme is expressed. An enzyme's activity is either a function, a terminal, or an output. Enzymes with terminal activities are called glands. Those with output activities are called receptors.

An enzyme has at least as many specificities as its activity has inputs. Each input is associated with one specificity, and this specificity determines which other enzyme this input will receive output from. In biological terms, a specificity determines which enzyme the input will recognize and bind as a substrate. Furthermore, specificity is defined in terms of shape, specifying the shape of the enzyme it would ideally bind. Note that, unlike in biology, specificity is given directly upon other enzymes rather than indirectly upon their products. This encourages product–substrate linkage to evolve, but without the overhead of maintaining data as a separate entity during execution.

An enzyme's shape determines how it is seen by other enzymes. A shape is a pattern that both identifies an enzyme and describes its role within the system. This latter role compares to biology, where shape and function are also related; though in biology function is determined by shape whereas here, conversely, shape is determined by function. A shape is a vector within a unitary n-dimensional space. However, before its derivation can be discussed, it is first necessary to understand how a collection of enzymes interact to form an executional structure.

Development

Development is the process that maps from an enzyme system's genome to the executional structure that it represents. In enzyme GP, an executional structure develops from the interactions of the components described within the genome. The interactions that occur are decided by a deterministic constraint satisfaction process, described below. From a biological perspective, this process resembles the minimization of binding energies in a system of protein-binding proteins and attempts to capture a generalized model of the construction of metabolic pathways within a cell.

To generate a valid executional structure, the hard constraints of development are that the structure contains a full set of outputs (receptors) and that each component (enzyme) which occurs within the structure has a full set of input sources. Further hard constraints may be given by the problem domain. For example, combinational logic design, the domain used to evaluate enzyme GP, requires executional structures be non-recursive. Given that the hard constraints are met, the aim of development is then to satisfy the input preferences of each enzyme of the executional structure; so that a component's actual input sources are as close as possible to the preferred input sources described by its specificities.

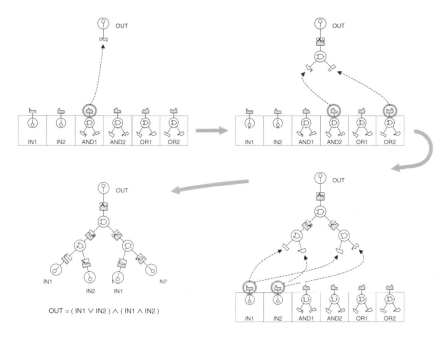

Figure 3.5 Development of a logic circuit from a simple genotype. The output terminal receptor binds an AND1 enzyme as its substrate. The AND1 enzyme now chooses its own substrates, and the process continues until the inputs of all expressed enzymes have been satisfied. Note that OR1 is never expressed, and IN1 and IN2 are both expressed twice.

There are potentially many ways of implementing this process. The implementation used in this study is a simple breadth-first search. Development begins with expression of the receptors, which then choose substrates whose shapes are most similar to their specificities. Substrates are chosen from those defined in the genotype and can be either enzymes or glands. These substrates are now considered expressed and, if they require inputs, attempt to satisfy them by binding their own substrates. This process continues in hierarchical fashion until all expressed receptors and enzymes have satisfied all of their inputs. An example of the process is illustrated in Figure 3.5 for the development of a combinational logic circuit. It is interesting to note that development does not require the executional structure to contain all the components described within the genome. Also, enzymes that are expressed may be used as input sources for more than one other enzyme.

Functionality as Shape

Recall that one of the problems with the parse tree representation of conventional GP is that context, which is determined by position, is not preserved by subtree crossover. In enzyme GP, the position of an enzyme within the genome has no effect on its context within the developed program. Its context is a result of the enzymes that choose it as an input source and the enzymes it chooses as input sources. Whether an enzyme is chosen as an input source at a particular location depends on its shape. Consequently, an enzyme's shape determines its context.

In biology, it is common for a molecule's shape to reflect (in part) its function. In enzyme GP, it is desirable that an enzyme's shape reflects its role within a program, for this would entail that enzymes with similar shapes would have similar roles. As such, during specificity satisfaction when an enzyme is selected for input according to its similarity to the specificity, the chosen enzyme will have a role similar to the preferred enzyme simply by definition of the process. If this were not the case, then the context of the enzyme—which depends on its inputs—would change considerably depending on which other enzymes were present within the program.

Functionality is a definition of shape, where an enzyme's shape is derived from its own activity and the activities of enzymes bound below it in the program. The aim of functionality is to make an enzyme's context meaningful and repeatable regardless of which other enzymes are present within a program.

A functionality is a point within functionality space. The easiest way to think of functionality space is as an enzyme reference system where each location characterizes a certain type of enzyme. Any enzyme can be referenced within this space and, consequently, other enzymes can use it to specify the types of enzymes they would prefer to receive input from.

Functionality space is a unitary vector space with a dimension for each member of the set of available functions and terminals. A functionality, therefore, is a vector with a component between 0 and 1 for each member of this set. A simple functionality space is depicted in Figure 3.6. The functionality, F, of a particular enzyme is derived using the following equations:

$$F(enzyme) = (1 - inputbias)F(activity) + inputbias.F(specificities) \quad (1)$$

$$F(specificities) = \frac{\sum_{i=1}^{n} specificity_i.strength(specificity_i)}{\sum_{i=1}^{n} strength(specificity_i)}, \quad (2)$$

where $inputbias$ is a constant and n is the number of specificities. Equation 1 states that the functionality of an enzyme is a weighted sum of the functionality of its activity and the functionality of its specificities. The functionality of the enzyme's activity, $F(activity)$, is a unit vector situated on the axis corresponding to the enzyme's function. A specificity is defined by the functionality it would like to match. The functionality of all the enzyme's specificities, defined in Equation 2, is a normalized sum of each specificity weighted by its strength. An illustrative example using these equations is shown in Figure 3.7.

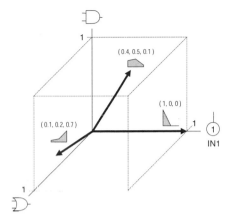

Figure 3.6 Functionality space for function set {AND, OR} and terminal set {IN1} showing example functionalities. Vector plots of functionalities are used for illustrative purposes in this chapter.

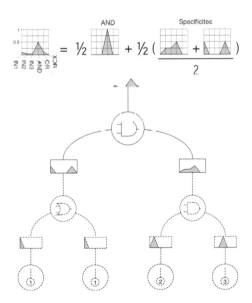

Figure 3.7 Derivation of functionality. An AND enzyme's functionality is derived from the functionality of the AND function and its specificities using Equations 1 and 2 with *inputbias* $= \frac{1}{2}$. The enzyme's functionality captures the function and terminal content, weighted by depth, of its ideal subtrees (shown dashed).

According to this definition of functionality, an enzyme's shape is derived from both its own activity and the shape of its specificities. The shapes of an enzyme's active specificities, moreover, are also the shapes of the enzymes that this enzyme would prefer to bind during development, its ideal substrates. Consequently, this enzyme's functionality also captures in part the functionalities of its ideal substrates, which also capture in part the functionalities of their ideal substrates. Following this logic recursively, it becomes evident that an enzyme's functionality captures, in part, the functionality of its ideal subtrees. Since functionality is also derived from activity, the functionality of an ideal subtree implicitly captures information regarding the occurrence of each activity within the subtree.

However, functionality only captures a profile of the functions and terminals, weighted by depth, within an enzyme's ideal subtrees. It does not capture the hierarchical structure of the trees. Consequently, a functionality cannot describe an enzyme uniquely. Nevertheless, functionality space is continuous, making it unlikely that two nonidentical enzymes will both have the same functionality and occur in the same program.

An enzyme's actual context depends on which other enzymes are present within a program. This implies that it is both variant between programs and indeterminate before development takes place. For both of these reasons, functionality cannot, and does not, attempt to specify an enzyme's actual context. It does, however, attempt to specify an enzyme's preferred context, which is the most probable context, but like any specific context, is itself unlikely to occur frequently.

Evolution

A program's genome, shown in Figure 3.8, is a linear array of enzymes grouped into subarrays of glands, functional enzymes, and receptors. In enzyme GP, a population of these genomes is evolved using a genetic algorithm with mutation and crossover.

The genetic algorithm, depicted in Figure 3.9, uses a spatially distributed population with a toroidal surface. The algorithm is generational, and during each generation every member of the population is mated with a copy of its fittest neighbor. The offspring are then subjected to mutation and fitness evaluation. If the fitness of a child solution is equal to or higher than the fitness of the existing solution at this location, then it replaces the existing solution. This is a locally elitist policy that does not allow fit solutions to be lost, but at the same time does not suffer the drawbacks of global elitism. For further details regarding this GA, see Lones and Tyrrell [23].

Enzyme GP uses two forms of crossover. The first, a uniform crossover, loosely models the mechanism of biological reproduction. The second, called

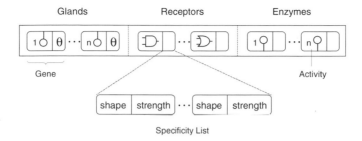

Figure 3.8 Genotype layout. The genotype defines which enzymes are present in an enzyme system. All genotypes declare a full set of terminal enzymes (glands and receptors). Note that glands do not have specificities, for they receive no input from other enzymes.

transfer and remove, attempts to capture some of the evolutionary behaviors induced by biological reproduction but does not use a biomimetic mechanism.

Uniform crossover is a two-stage process of gene recombination followed by gene shuffling (Figure 3.10). Gene recombination resembles meiosis, the biological process whereby maternal and paternal DNA is recombined to produce germ cells. Gene recombination involves selecting and recombining a number of pairs of similar genes from the parent genomes. Pairs of genes are selected according to the similarity of the functionalities of their enzyme products. Recombination involves specificity recombination where pairs of functionalities are recombined using standard GA uniform crossover, followed by specificity shuffling, where the recombined specificities are divided uniformly between the child genes. Gene shuffling then divides the recombined genes uniformly between the child genotypes.

A general view of recombination in natural systems is that it changes the genetic makeup of a parent genome by adding some new genes and removing some existing genes. Transfer and remove (TR) crossover is based on this idea. A TR crossover event generates a child solution by either copying a contiguous

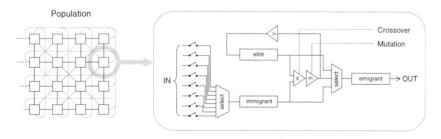

Figure 3.9 Genetic algorithm structure, showing the processing that occurs in each cell of the distributed population during each generation.

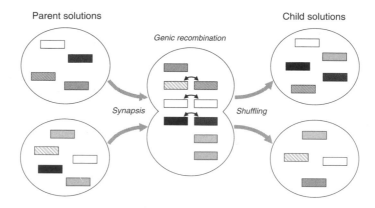

Figure 3.10 A conceptual view of uniform crossover. Genes are shown as rectangles. Shade indicates functionality.

group of genes from one parent to another or by removing a contiguous group of genes from a parent (Figure 3.11). Each of these operations is applied in 50% of recombination events. In subtree crossover, one subtree is always replaced by another. Using TR recombination, components being added and components being removed from a program are independent events, and therefore they are less disruptive than using subtree or uniform crossover. Furthermore, the components (apart from receptors) transferred to a program will only become involved in the program's execution if they fulfill the interaction preferences of existing components. Consequently, new components will only replace existing components if they fulfill these preferences better than the components they replace.

The mutation operator of enzyme GP targets both activities and specificities. An activity mutation replaces the current activity with one chosen randomly

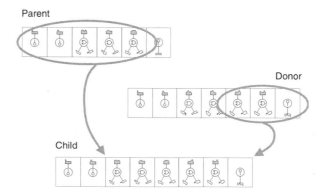

Figure 3.11 Recombination using transfer and remove. Note that the number of receptors in a genotype is fixed, so any receptors transferred from the donor replace those copied from the parent.

Table 3.1 Test problems.

Name	Inputs	Outputs	Function set
Two-bit full adder	5	3	XOR, MUX
Two-bit multiplier	4	4	AND, XOR
Even-three-parity	3	1	AND, OR, NAND, NOR
Even-four-parity	4	1	AND, OR, NAND, NOR

from the function or terminal sets. A specificity mutation can change both dimensions of functionalities and specificity strengths, replacing the old value with a randomly chosen new value.

DIGITAL CIRCUIT DESIGN

Enzyme GP has been applied to a number of problems in the domain of combinational logic design—the design of non-recurrent digital circuits. This section discusses the performance of enzyme GP on the set of problems listed in Table 3.1. Parameter settings are recorded in Table 3.2. Results for enzyme GP using uniform crossover, TR crossover, and no crossover are shown in Table 3.3. Computational effort [16] measures the number of evaluations required for a 99% confidence of finding an optimal solution.

Table 3.3 shows that, in all but the even-four-parity problem, enzyme GP with crossover performs better than enzyme GP with mutation alone. It also shows that the computational effort using uniform crossover is less than using TR crossover. Nevertheless, it is interesting to note that although uniform crossover generally finds a solution in a higher proportion of runs, TR crossover generally finds a solution earlier in a successful run. There are two apparent reasons that uniform crossover has a higher success rate. First, genome length is bounded so that all possible lengths are sufficient to contain an optimal solution. For TR crossover this is not the case and, whereas solutions start with a viable length, TR crossover is free to explore solutions that are shorter than the minimum

Table 3.2 Parameter settings.

Parameter	Value
Input bias (see Equation 1)	0.3
Number of specificities per enzyme	3
Distance limit for gene alignment in uniform c/o	1
Proportion of gene pairs recombined	15%
Transfer size range during TR crossover	1–5
Rate of specificity strength mutation	2%
Rate of functionality dimension mutation	15%

Table 3.3 Performance of enzyme GP with different operators.

Operator	Average	Success	CE
Two-bit multiplier, bounds 12–16, pop. 324			
Uniform	118	77%	136,080
TR	76	75%	169,128
Mutation	94	56%	334,368
Two-bit adder, bounds 10–20, pop. 324			
Uniform	113	74%	244,620
TR	107	57%	340,200
Mutation	114	53%	392,364
Even-three-parity, bounds 5–10, pop. 100			
Uniform	54	43%	79,000
TR	47	32%	96,000
Mutation	40	7%	250,800
Even-four-parity, bounds 10–25, pop. 625			
Uniform	150	25%	2,588,750
TR	165	24%	2,703,125
Mutation	154	36%	1,830,000

viable length. Second, uniform crossover is more disruptive than TR crossover because it targets multiple sections of the genome and always causes genes to be replaced rather than appended. Consequently, it is able to explore a larger region of the search space during a run. Conversely, TR crossover prefers exploitation of existing genetic material to exploration of new material and, accordingly, is perhaps able to carry out a more structured search and reach the optimum in less generations.

The adder, multiplier, and even-four-parity problems have been solved by Miller [29] using Cartesian GP [28], a graph-based genetic programming system. Results from Koza [15, 16], using tree-based GP, are available for both parity problems. Koza [16] has also attempted the 2-bit adder problem, but quantitative results are not available. Miller records minimum computational effort of between 210,015 and 585,045 for the 2-bit multiplier problem. Enzyme GP, requiring a minimum computational effort of 136,080 for a population of 324, compares favorably with these results. For the 2-bit adder problem, Miller cites a minimum computational effort of 385,110. For enzyme GP, using the same functions as Miller, mimimum effort is 244,620.

Koza has evolved even-*n*-parity circuits using populations of size 4000 [16] and 16,000 [15]. For the even-three-parity problem (and without using Automatically Defined Functions [ADFs]), this gives minimum computational efforts of 80,000 and 96,000 respectively. For the even-four-parity problem, minimum computational efforts are 1,276,000 and 384,000, respectively. For enzyme GP, minimum computational effort has been calculated at 79,000 for the even-three-parity problem with a population of 100. For the even-four-parity problem with a population of 625, minimum computational effort is 2,588,750 with crossover and 1,830,000 without. This suggests that enzyme GP cannot easily evolve even-n-parity circuits where $n > 3$, at least for these (relatively small) population sizes, and especially when crossover is used. This agrees with Miller's findings, where only 15 correct even-four-parity circuits were found in the course of 400 million evaluations. Langdon [18] suggested that parity circuits are unsuitable benchmarks for GP. Furthermore, parity circuits involve a high degree of structure reuse, and it seems possible that subtree crossover, which is able to transfer substructures independently of their original context, may be more able to develop this sort of pattern than enzyme GP.

SOLUTION SIZE EVOLUTION

Another problem with GP is bloat, the phenomenon whereby program size increases considerably faster than program functionality. In standard GP, program size growth is near quadratic [19], yet the exact causes of bloat are not known. Nevertheless, a number of theories have been proposed. These include hitchhiking [39], protection from disruptive operators [7], operator biases [1], removal biases [38], and search space bias [21].

Enzyme GP, in contrast, does not suffer from bloat [24, 25]. This is true both when solution length is bounded, which is the case for uniform crossover, and when length is unbounded, in the case of TR crossover. Genome size evolution for both operators is shown in Figure 3.12. These graphs show three different scenarios. In the top-left graph, initial phenotype length is below the minimum length required to solve the problem. In the bottom-left graph, the initial length is above and near the minimum length. In the graphs on the right, initial length is well above the minimum length.

Where the initial phenotype size is below the minimum size of an optimal solution, there is an average increase in both genotype and phenotype size, with phenotype size growing faster than genotype size and both leveling off once the phenotype is large enough to solve the problem. Where phenotype size is near and above the minimum, average phenotype size remains constant, though genotype size still demonstrates a net growth. Where phenotype size is well above the minimum, there is an average decrease in phenotype size, whereas genotype size remains, on average, constant.

Interestingly, this suggests a bias in phenotype size evolution toward short, yet viable, solution sizes. At the same time, there is no bias in genotype size so long as it is sufficient to contain the phenotype. Enzyme GP does not penalize longer solutions during evaluation and, consequently, there must be an

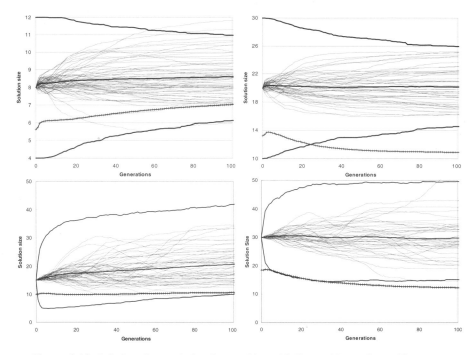

Figure 3.12 Solution size evolution for two-bit multiplier problem using uniform crossover (top) and TR crossover (bottom). Light lines show average genotype sizes for each run. Dark uncrossed lines show average minimum, average, and maximum solution sizes across all runs. Dark crossed line shows average phenotype size across all runs. Minimum optimal phenotype length is seven gates.

inherent evolutionary advantage for smaller solutions. The reason for this is not yet known, though a speculative explanation is that larger solutions are less stable than smaller solutions and therefore less able to accurately replicate during recombination. This instability might lie in a greater presence of poor specificity-shape matches in large solutions, which, during recombination, are more likely to be broken and reformed into other interactions. Nevertheless, in one sense it is an advantage that enzyme GP automatically searches for short solutions, yet equally, it suggests that large solutions to large problems might be harder to find.

DISCUSSION

Conceptually, a conventional parse tree can be transformed into an enzyme GP program by decomposing it into its functional components and translating each component's context, previously given by its position within the tree, into a description of the components it was connected to in the tree. The components, each with its associated context, are then stored in a linear array.

This new way of representing the parse tree has a number of advantages from an evolutionary perspective. First, it introduces positional independence. The position of a component in the array need not have any direct bearing on the position of the component in the parse tree. This is of benefit to recombination because position is not conserved where different programs are of different shapes and sizes—which, most of the time, is the case for GP. Consequently, the position of a component in a child program will most likely not be its position in the parent program. In tree-based GP, where context is given by position, this is a problem. In enzyme GP, where context is independent of position, it is not a problem.

Second, the linear encoding has a homogenous structure. In a typical parse tree about half of the nodes are leaf nodes. Consequently, during subtree crossover, a large proportion of randomly chosen crossover points will occur at or near the leaf nodes. This, in turn, means that a disproportionately high number of small subtrees will be targeted during crossover. In practice the selection of crossover points is biased so that more are chosen nearer the root of the tree. However, this bias has to be carefully chosen to avoid biasing the shapes and sizes of solutions explored during evolution. In enzyme GP, by comparison, there is no need to bias crossover because it will automatically sample subtrees of all sizes. Furthermore, enzyme GP does not have to target complete subtrees during recombination. It can, in principle, target any collection of components. In some cases, components will be connected, such as subtrees, in other cases they will not. Consequently, crossover can carry out a wide range of behaviors. However, it is possible that during evolution linkage will develop between neighboring genes, meaning that crossover is more likely to target connected groups of components. The more evolved the solutions, therefore, the more focused the behaviors of crossover could become.

Third, not all the components described within a genome need be expressed in the developed program. One example of the utility of this is TR crossover, which allows components to be added to a genome without replacing existing components. If the new components do not naturally interact with the existing components, then they will not be used in the program. This is less disruptive than forcing components to be replaced, which is the only option with subtree crossover. Redundant components, more generally, can form a source of backtracking and neutral evolution. Backtracking can be enabled by saved copies of components that are no longer used in developed programs. These, and unused copies of current components, are subject to mutation. Because they do not contribute to the phenotype, these mutations are neutral and have no fitness penalty. Nevertheless, these mutations may be beneficial and result in improvements to existing components. This might lead to the development of gene families comparable to those found in biological genomes.

CONCLUSIONS

This chapter presented an overview of enzyme genetic programming, an approach to evolutionary computation motivated by the metabolic processing of cells. Enzyme GP can be distinguished from conventional GP by its use of a program representation and developmental process derived from biology. The aim of the approach is to capture the elements of biological representations that contribute to their evolvability and adapt these for artificial evolution. The resulting system represents an executable structure as a collection of enzymelike computational elements that interact with one another according to their own interaction preferences.

The method has been applied to a number of problems in the domain of digital circuit design but does not yet indicate any significant performance advantage over other GP approaches. However, analysis of solution-size evolution shows that, unlike most other GP approaches, enzyme GP does not suffer from bloat. On the contrary, enzyme GP appears to be biased toward finding smaller solutions to problems, and it achieves this without any form of fitness penalty or operator modification.

Enzyme GP has a number of interesting properties. Perhaps most important of these is that the context of each component within a program is independent of its position within the genome. Furthermore, the context of a component is recorded using a description independent of any particular program and, consequently, the role of a component can be preserved following recombination.

References

[1] L. Altenberg. Emergent phenomena in genetic programming. In A. V. Sebald and L. J. Fogel, editors, *Evolutionary Programming—Proceedings of the Third Annual Conference*, pages 233–241. World Scientific Publishing, River Edge, NJ, 1994.

[2] Lee Altenberg. The evolution of evolvability in genetic programming. In K. Kinnear, Jr., editor, *Advances in Genetic Programming*, pages 47–74. MIT Press, Cambridge, 1994.

[3] P. Angeline. Subtree crossover: building block engine or macromutation? In J. Koza, editor, *Genetic Programming 1997: Proceedings of the Second Annual Conference, GP97*, pages 240–248. Morgan Kaufmann, San Francisco, CA, 1997.

[4] P. Angeline. Multiple interacting programs: A representation for evolving complex behaviors. *Cybernet. Syst.*, 29(8):779–806, 1998.

[5] T. Bäck. *Evolutionary Algorithms in Theory and Practice.* Oxford University Press, Oxford, 1996.

[6] Wolfgang Banzhaf, Peter Nordin, Robert E.Keller, and Frank D. Francone. *Genetic Programming—An Introduction.* Morgan Kaufmann, San Francisco, CA, 1998.

[7] T. Blickle and L. Thiele. Genetic programming and redundancy. In J. Hopf, editor, *Genetic Algorithms within the Framework of Evolutionary Computation (Work-*

shop at KI-94, Saarbrücken), pages 33–38. Max-Planck-Institut für Informatik, Saarbrücken, Germany, 1994.

[8] D. Bray. Protein molecules as computational elements in living cells. *Nature*, 376:307–312, 1995.

[9] M. Capstick, W. P. Liam Marnane, and R. Pethig. Biological computational building blocks. *IEEE Computer*, 25(11):22–29, 1992.

[10] M. Conrad. Molecular computing: the lock-key paradigm. *IEEE Computer*, 25(11): 11–20, 1992.

[11] M. J. Fisher, R. C. Paton, and K. Matsuno. Intracellular signalling proteins as 'smart' agents in parallel distributed processes. *BioSystems*, 50:159–171, 1999.

[12] L. Fogel, A. Owens, and M Walsh. *Artificial Intelligence through Simulated Evolution*. Wiley, New York, 1966.

[13] D. E. Goldberg. *Genetic Algorithms in Search, Optimisation and Machine Learning*. Addison-Wesley, Reading, MA, 1989.

[14] Holland J. II. *Adaptation in Natural and Artificial Systems*. MIT Press, Cambridge, 1975.

[15] J. Koza. *Genetic Programming II: Automatic Discovery of Reusable Programs*. MIT Press, Cambridge, 1994.

[16] John Koza. *Genetic Programming: On the Programming of Computers by Means of Natural Selection*. MIT Press, Cambridge, 1992.

[17] John Koza, Forrest Bennett, III, David Andre, and Martin Keane. *Genetic Programming III: Darwinian Invention and Problem Solving*. Morgan Kaufmann, San Francisco, CA, 1999.

[18] W. Langdon and R. Poli. Why "building blocks" don't work on parity problems. Technical Report CSRP-98-17, School of Computer Science, University of Birmingham, UK, July 1998.

[19] W. B. Langdon. Quadratic bloat in genetic programming. In D. Whitley, editor, *Proceedings of the 2000 Genetic and Evolutionary Computation Conference*, pages 451–458. Morgan Kaufmann, San Francisco, CA, 2000.

[20] W. B. Langdon. Size fair and homologous tree genetic programming crossovers. *Genet. Program. Evolvable Machines*, 1(1/2):95–119, April 2000.

[21] W. B. Langdon and R. Poli. Fitness causes bloat. In P. K. Chawdhry, editor, *Soft Computing in Engineering Design and Manufacturing*, pages 13–22. Springer, Berlin, 1997.

[22] M. A. Lones and A. M. Tyrrell. Biomimetic representation in genetic programming. In H. Kargupta, editor, *Proceedings of the 2001 Genetic and Evolutionary Computation Conference Workshop Program*, pages 199–204, IEEE Press, Piscataway, NJ, July 2001.

[23] M. A. Lones and A. M. Tyrrell. Enzyme genetic programming. In *Proceedings of the 2001 Congress on Evolutionary Computation*, vol. 2, pages 1183–1190. IEEE Press, Piscataway, NJ, May 2001.

[24] Michael A. Lones and Andy M. Tyrrell. Biomimetic representation with enzyme genetic programming. *J. Genet. Program. Evolvable Machines*, 3(2):193–217, 2002.

[25] Michael A. Lones and Andy M. Tyrrell. Crossover and bloat in the functionality

model of enzyme genetic programming. In D. B. Fogel, M. A. El-Sharkani, X. Yao, G. Greenwood, H. Iba, P. Marrow and M. Shackleton, editors, *Proceedings of the Congress on Evolutionary Computation*, pages 986–991. IEEE Press, Piscataway, NJ, 2002.

[26] S. Luke, S. Hamahashi, and H. Kitano. "Genetic" programming. In W. Banzhaf, editor, *GECCO-99: Proceedings of the Genetic and Evolutionary Computation Conference*, pages 1098–1105. Morgan Kaufmann, San Francisco, CA, 1999.

[27] P. C. Marijuán. Enzymes, artificial cells and the nature of biological information. *BioSystems*, 35:167–170, 1995.

[28] J. Miller and P. Thomson. Cartesian genetic programming. In R. Poli, editor, *Third European Conference on Genetic Programming*, vol. 1802 of *Lecture Notes in Computer Science*, pages 121–132. Springer, Berlin, 2000.

[29] J. F. Miller, D. Job, and V. K. Vassilev. Principles in the evolutionary design of digital circuits—part I. *Genet. Program. Evolvable Machines*, 1:7–36, 2000.

[30] P. Nordin and W. Banzhaf. Complexity compression and evolution. In L. Eshelman, editor, *Genetic Algorithms: Proceedings of the Sixth International Conference (ICGA95)*, pages 310–317. Morgan Kaufmann, San Francisco, CA, 1995.

[31] P. Nordin, F. Francone, and W. Banzhaf. Explicitly defined introns and destructive crossover in genetic programming. In P. Angeline and K. Kinnear, Jr., editors, *Advances in Genetic Programming 2*, pages 111–134. MIT Press, Cambridge, 1996.

[32] R. Poli. Evolution of graph-like programs with parallel distributed genetic programming. In E. Goodman, editor, *Proceedings of Seventh International Conference on Genetic Algorithms*, pages 346–353. Morgan Kaufmann, San Francisco, CA, 1997.

[33] I. Rechenberg. *Evolutionstrategie: Optimierung Technisher Systeme nach Prinzipien der Biologischen Evolution*. Frommann-Hoolzboog Verlag, 1973.

[34] V. Reddy, M. Mavrovouniotis, and M. Liebman. Petri net representations in metabolic pathways. In L. Hunter, editor, *Proceedings of the First International Conference on Intelligent Systems for Molecular Biology*. MIT Press, Cambridge, 1993.

[35] C. Ryan, J. J. Collins, and M. O'Neill. Grammatical evolution: evolving programs for an arbitrary language. In W. Banzhaf, editor, *First European Workshop on Genetic Programming*, vol. 1391 of *Lecture Notes in Computer Science*. Springer, Berlin, 1998.

[36] H. Schwefel. Kybernetische evolution als strategie der experimentallen forschung in der strömungstechnik. Diplomarbeit, 1965.

[37] M. Shackleton and C. Winter. A computational architecture based on cellular processing. In M. Holcombe and R. Paton, editors, *Proceedings of the International Conference on Information Processing in Cells and Tissues (IPCAT)*, pages 261–272. Plenum Press, New York, 1997.

[38] T. Soule, J. A. Foster, and J. Dickinson. Code growth in genetic programming. In J. R. Koza, editor, *Genetic Programming 1996: Proceedings of the First Annual Conference*, pages 215–213. MIT Press, Cambridge, 1996.

[39] W. A. Tackett. *Recombination, Selection, and the Genetic Construction of Computer Programs*. PhD thesis, University of Southern California, 1994.

4

Genetic Process Engineering

Ron Weiss, Thomas F. Knight Jr.,
and Gerald Sussman

In this chapter we present an engineering discipline to obtain complex, predictable, and reliable cell behaviors by embedding biochemical logic circuits and programmed intercellular communications into cells. To accomplish this goal, we provide a well-characterized component library, a biocircuit design methodology, and software design tools.

Using the cellular gates, we introduce *genetic process engineering*, a methodology for modifying the DNA encoding of existing genetic elements to achieve the desired input/output behavior for constructing reliable circuits of significant complexity. We also describe BioSpice, a prototype software tool for biocircuit design that supports both static and dynamic simulations and analysis of single-cell environments and small cell aggregates.

INTRODUCTION

The goal of our research is to lay the foundations of an engineering discipline for building novel living systems with well-defined purposes and behaviors using standardized, well-characterized components. Cells are miniature, energy efficient, self-reproduce, and can manufacture biochemical products. These unique characteristics make cells attractive for many novel applications that require precise programmed control over the behavior of the cells. The applications include nanoscale fabrication, embedded intelligence in materials, sensor/effector arrays, patterned biomaterial manufacturing, improved

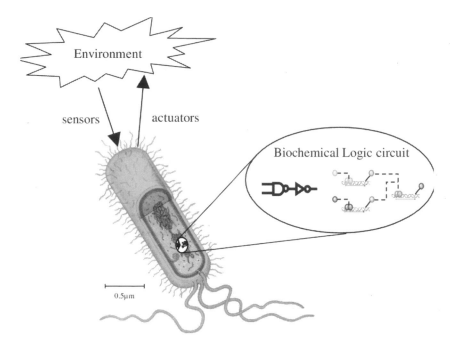

Figure 4.1 Embedding biochemical logic circuits in cells for internal computation and programmed intercellular communications, extending and modifying the behavior of cells and cell aggregates.

pharmaceutical synthesis, programmed therapeutics, and as a sophisticated tool for *in vivo* studies of genetic regulatory networks. These applications require synthesis of sophisticated and reliable cell behaviors that instruct cells to make logic decisions based on factors such as existing environmental conditions and current cell state. For example, a cell may be programmed to secrete particular timed sequences of biochemicals depending on the type of messages sent by its neighbors. The approach proposed here for engineering the requisite precision control is to embed internal computation and programmed intercellular communications into the cells (Figure 4.1). The challenge is to provide robust computation and communications using a substrate where reliability and reproducible results are difficult to achieve.

Biological organisms as an engineering substrate are currently difficult to modify and control because of the poor understanding of their complexity. Genetic modifications to cells often result in unpredictable and unreliable behavior. A single *Escherichia coli* bacterial cell contains approximately 10^{10} active molecules, about 10^7 of which are protein molecules. The cumulative interaction of these molecules with each other and the environment determines the behavior of the single cell. Although complex, these interactions are not arbitrary. Rather, cells are highly optimized information-processing units that monitor their environment and continuously make decisions on how to react to the given conditions. Moreover, from prokaryotes to eukaryotes, cells act both as individual units and as a contributing part of larger and complex multicellular systems or organisms. Given these attributes and inherent complexity, how can we successfully modify and harness biological organisms for our purposes?

Controlled gene expression using engineered *in vivo* digital-logic circuits and intercellular communications enables programmed cell behavior that is complex, predictable, and reliable. Our approach integrates several layers that include a library of well-characterized simple components synthesized to have the appropriate behavior, a methodology for combining these components into complex intracellular circuitry and multicellular systems with predictable and reliable behavior, and software tools for design and analysis.

The first step in making programmed cell behavior a practical and useful engineering discipline is to assemble a component library. For this purpose, we engineered cellular gates that implement the NOT, IMPLIES, and AND logic functions. These gates are then combined into biochemical logic circuits for both intracellular computation and intercellular communications. In these biocircuits, chemical concentrations of specific messenger RNA (mRNA) and inducer molecules represent the logic signals. The logic gates perform computation and communications using mRNA, DNA-binding proteins, small inducer molecules that interact with these proteins, and segments of DNA that regulate the expression of the proteins. For example, Figure 4.2 describes how a cellular inverter achieves the two states in digital inversion using these genetic regulatory elements.

Given a library of components, biocircuit design is the process of assembling preexisting components into logic circuits that implement specific behaviors. The most important element of biocircuit design is matching logic gates such that the couplings produce the correct behavior. Typically, naturally occurring components have widely varying kinetic characteristics, and arbitrarily composing them into circuits is not likely to work. We demonstrate *genetic process engineering*—modifying the DNA encoding of existing genetic elements until they achieve the desired behavior for constructing reliable circuits of significant complexity. The genetic modifications produce components that implement digital computation with good noise margins, signal restoration, and appropriate standard interfaces for complex system composition.

An important aspect of this work is engineering biological systems to exhibit digital behavior because the digital abstraction is both convenient to use and feasible. The digital abstraction is a useful programming paradigm because it offers a reliable and conceptually simple methodology for constructing complex behavior from a small number of simple components [18]. Digital computation provides reliability by reducing the noise in the system through signal restoration. For each component in the computation, the analog output

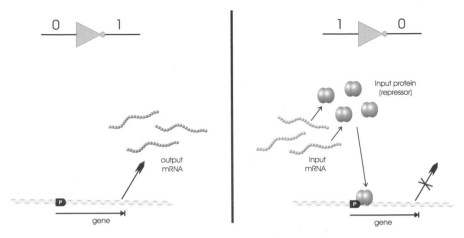

Figure 4.2 A simplified view of the two cases for a biochemical inverter. Here, the concentration of a particular messenger RNA (mRNA) molecule represents a logic signal. In the first case, the input mRNA is absent and the cell transcribes the gene for the output mRNA. In the second case, the input mRNA is present and the cell translates the input mRNA into the input protein. The input protein then binds specifically to the gene at the promoter site (labeled P) and prevents the cell from synthesizing the output mRNA.

signal represents the digital value better than the analog input signal. Engineers carefully combine these reliable components into complex systems that perform reliable computation. Experimental results in chapter 7 describe *in vivo* digital-logic circuits with good noise margins and signal restoration to demonstrate the feasibility of programming cells using digital computation. In the study of existing biological organisms, recent work [7–9, 13] suggests that cells routinely use digital computation to make certain decisions that result in binary on/off behavior. Because the digital abstraction is both convenient to use and feasible, it offers a useful paradigm for programming cells and cell aggregates. However, much like desktop computers use both digital and analog components, in the future we will also incorporate analog logic elements into engineered biological systems as the analog components become necessary for particular tasks.

To create novel biological systems, an engineer must be equipped with design and modeling software that prescribes how primitive components may be combined into complex systems with predictable and reproducible behavior. We present BioSPICE, a prototype tool that helps biocircuit designers manage the complexity of the substrate and achieve reliable systems. The inspiration for BioSPICE comes from the utility of tools such as SPICE [10] in the design of electrical circuits. BioSPICE supports both static and dynamic analysis of single-cell environments and small cell aggregates.

For obtaining coordinated aggregate cell behavior, we demonstrate programmed cell-to-cell communications using chemical diffusion to carry messages. Multicellular organisms create complex patterned structures from identical, unreliable components. This process of differentiation relies on communications between the cells that compose the system. Learning how to engineer such robust behavior from aggregates is important for better understanding distributed computing, for better understanding the natural developmental process, and for engineering novel multicellular organisms with well-defined behaviors. Chemical diffusion is one of several communication mechanisms that can help achieve coordinated behavior in cell aggregates.

In this chapter, the next section introduces the logic gates we implemented in cells. The NOT gate is the fundamental building block for constructing intracellular circuits, and the IMPLIES and AND gates are used for intercellular communications. The third section introduces the BioSPICE tool for biocircuit design and analysis. The section describes the model used for simulating a biochemical inverter, simulations of simple logic circuits in single cells, and analysis of genetic modifications to achieve the desired gate behavior.

CELLULAR GATES

A fundamental chemical process in the cell is the production of proteins from genes encoded in the DNA. The cell performs important regulatory activities through DNA-binding proteins that repress or activate the production of specific proteins. Repression can be used to implement digital-logic inverters [6]. This section presents a formal model of this inversion mechanism and explains how to construct any finite digital-logic circuit using these inverters.

In this section we also introduce two additional logic gates that implement the IMPLIES and AND logic functions. The inputs to these gates are mRNA molecules that code for DNA-binding proteins and small inducer molecules that affect the activity of these proteins. Because the inducer molecules freely diffuse through cell membranes, these two gates are useful for intercellular communications and other external interactions with the *in vivo* circuits.

A Biochemical Inverter

Natural gene regulation systems exhibit characteristics useful for implementing *in vivo* logic circuits. Figure 4.2 presents a simplified view of the two states in the biochemical process of inversion in either the presence or absence of the input mRNA signal. Figure 4.3 shows a more detailed description of the inversion process, including the role of transcription and translation.

Figure 4.4 illustrates the functional construction of an inverter from its biochemical reaction stages. Let ψ_A denote the concentration level of the input

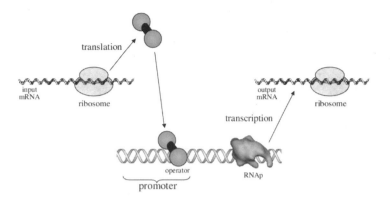

Figure 4.3 Biochemical inversion uses the transcription and translation cellular processes. Ribosomal RNA translates the input mRNA into an amino acid chain, which then folds into a three-dimensional protein structure. When the protein binds to an operator of the gene's promoter, it prevents transcription of the gene by RNA polymerase (RNAp). In the absence of the repressor protein, RNAp transcribes the gene into the output mRNA.

mRNA, representing the input signal to the inverter. In the first stage, ribosomal RNA translates the mRNA product ψ_A into the input repressor protein ϕ_A. Let $\mathcal{L}$ (translation stage) denote the steady-state mapping between ψ_A and ϕ_A. In general, increases in ψ_A yield linear increases in ϕ_A until an asymptotic boundary is reached. Factors that determine this boundary include the amino acid synthesis capabilities of the cell, the efficiency of the ribosome-binding site, and mRNA stability. Because the cell degrades both mRNA and input protein molecules, a continuous synthesis of the input mRNA is required for a steady level of the input protein.

The second stage in the inverter uses cooperative binding to reduce the digital noise. Here, the input protein monomers join to form polymers (often dimers, occasionally tetramers), which then bind to the operator and repress the gene. Because the cells degrades the proteins, a continuous synthesis of the input protein is required for maintaining the repression activity. Let ρ_A denote the strength of this repression, defined as the concentration of operator that is bound by repressor. In steady state, the relation $\mathcal{C}$ (cooperative binding stage) between ϕ_A and ρ_A will generally be sigmoidal. For low values of ϕ_A, the amount of repression increases only slightly as the input protein concentrations increase because these concentrations are too low for significant dimerization. Without dimerization, the monomeric repressor cannot bind to the DNA. Then, at higher levels of ϕ_A (when the input proteins dimerize easily), cooperative binding and dimerization result in nonlinear increases in the repression activity. Finally, at saturating levels of the input protein when the operator is mostly bound, the curve reaches an asymptotic boundary. Because the repressor activity is

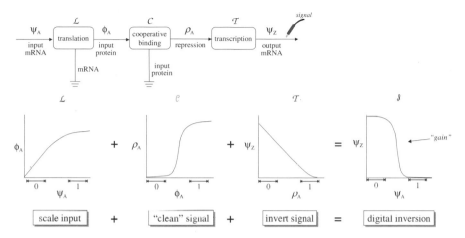

Figure 4.4 Functional composition of the inversion stages; the translation stage maps input mRNA levels (ψ_A) to input protein levels (ϕ_A), the cooperative binding stage maps input protein levels to bound operator levels (ρ_A), and the transcription stage maps bound operator levels to output mRNA levels (ψ_Z). The degradation of the mRNA and protein molecules is represented with the electrical ground symbol. The degradation of mRNA is part of the translation stage, while the degradation of proteins is part of the cooperative binding stage. The graphs illustrate the steady-state relationships for each of these stages and the overall inversion function that results from combining these stages.

maximal, additional repressor molecules have no noticeable effect. The purpose for this stage is to provide signal restoration: as a result of this stage, the analog input signal better approximates its digital meaning.

In the last stage, RNA polymerase transcribes the structural gene and inverts the signal. Let ψ_Z denote the mRNA concentration for the output signal Z. Then, in the steady-state relation T (transcription stage) between ψ_Z and ρ_A, ψ_Z decreases monotonically as ρ_A increases. With no repression, transcription proceeds at maximum pace (i.e., maximum level of ψ_Z). Any increase in repression activity results in a decrease in transcription activity, and hence the inversion of the signal.

As illustrated in Figure 4.4, the functional combination $\mathcal{I}$ of the above stages achieves digital inversion:

$$\psi_Z = \mathcal{I}(\psi_A) = T \circ C \circ \mathcal{L}(\psi_A).$$

$\mathcal{I}$ is the transfer function of the inverter.

Intracellular Logic Circuits

Biochemical inverters are the building blocks for constructing intracellular logic circuits. First, most protein-coding sequences can be successfully fused to any

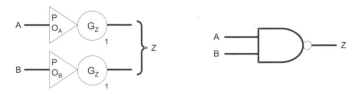

Figure 4.5 Wire OR-ing the outputs of two inverters yields a NAND gate.

given promoter. Second, two inverters form a NAND gate by "wiring-OR" their outputs (Figure 4.5). These two features combine to provide a modular approach to logic circuit design of any finite complexity, as described below.

Modularity in Circuit Construction

The modularity in biocircuit design stems from the ability to designate almost any gene as the output of any logic gate. Consider a logic element A consisting of an input mRNA, M_A, that is translated into an input protein repressor R_A, acting on an operator, O_A, associated with a promoter, P_A. Let P_A be fused to a structural gene, G_Z, coding for the output mRNA, M_Z. Figure 4.3 illustrates these genetic elements. The DNA basepair sequence G_Z (or the corresponding output mRNA sequence M_Z) that codes for an output protein, R_Z, determines the gate connectivity because the output protein may bind to other operators in the system. The specific binding of R_Z to another downstream operator, O_Z, connects gates because the level of R_Z affects the operation of the downstream gate.

To a first approximation, the choice of the sequence G_Z does not affect the transfer function, $\mathcal{T}$, of the inverter, $M_Z = \mathcal{T} \circ \mathcal{C} \circ \mathcal{L}(M_A)$. An exception to this rule occurs when G_Z codes for a protein that interacts with operator O_A or with input protein repressor R_A. Thus, the designer of *in vivo* logic circuits must ensure that the signal proteins do not interact with circuit elements other than their corresponding operators. The circuit designer should experimentally verify this required protein noninterference before circuit design.[1] Any set of non-interacting proteins can then serve as a library of potential signals for constructing an integrated circuit.

Once protein noninterference is established, modularity of the network design affords a free choice of signals. Any suitable repressor protein and its corresponding mRNA is a potential candidate for any signal, where the issue of suitability is discussed later in this chapter. This modularity is necessary for implementing a *biocompiler*—a program that consults a library of repressor

---

[1] For example, the following simple *in vivo* experiment checks whether a protein affects a particular promoter. First, fuse a fluorescent protein to a promoter of interest and quantify the *in vivo* fluorescence intensity. Next, add a genetic construct that overexpresses the protein of interest. Finally, check the new fluorescence intensity to determine whether the protein affects transcription.

proteins and their associated operators and generates genetic logic circuits directly from gate-level descriptions. Contrast this modularity with the method of Hjelmfelt et al. [5], which requires proteins that modify other proteins and where all signals are protein concentrations. In that case, the resulting physicochemical interdependence of successive logic stages makes simple modularity almost impossible.

Implementation of Combinatorial Logic

The approach to combinatorial logic is to "wire-OR" the outputs of multiple inverters by assigning them the same output gene. The output mRNA is expressed in the absence of either input mRNAs, and is not be expressed only when both inputs are present. This configuration implements a NAND gate. Because the performance of a NAND gate relies solely on that of its constituent inverters, well-engineered inverters will yield well-engineered combinatorial gates.

Figure 4.6 illustrates a small circuit where a NAND gate connects to an inverter. Here, mRNA and their corresponding proteins serve as the logic circuit wires, while the promoter and protein/mRNA decay implement the gates. Because a NAND gate is a universal logic element and can be wired to other gates, any finite digital circuit can be built within practical limitations such as the number of distinct signal proteins available.

Choice of Signals

The library of naturally available signal proteins includes approximately a few thousand candidates. Any repressor protein with sufficiently cooperative DNA binding that does not interfere with normal cell operation is a potential candidate. The experiments described in chapter 7 use four different naturally occurring DNA binding proteins and their operators, as well as several novel mutations to one of the operators. The number of naturally occurring proteins that could potentially serve as signals may never limit biocircuit construction because other factors, such as the metabolic capabilities of cells, are likely to place lower limits on biocircuit complexity within a single cell. However, it

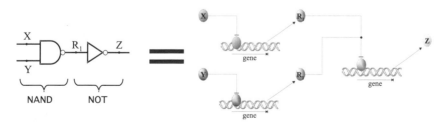

Figure 4.6 A logic circuit and its DNA implementation: The wire-OR of the outputs of two genes implements a NAND gate, and the choice of the output gene determines the gate connectivity.

may still be more efficient to synthesize artificial DNA-binding proteins for use as signals rather than finding natural sources. In the future, combinatorial chemistry techniques, along with a method such as phage display, will yield large libraries of novel DNA-binding proteins and corresponding operators. One potential source of a very large set of noninteracting signals is engineered zinc finger DNA binding proteins [3].

Intercellular Gates

Although the biochemical inversion mechanism suffices for building intra-cellular circuits, external interaction with the cells requires additional logic gates. Small molecules known as inducers freely diffuse through cellular membranes and interact with DNA-binding proteins. This section describes how the inducer-protein interactions implement two different intercellular gates. Chapter 7 reports on experimental results where the IMPLIES gate enables human-to-cell communications, and the AND gate facilitates cell-to-cell communications.

The IMPLIES Gate

The IMPLIES gate allows cells to receive control messages sent by humans or to detect certain environmental conditions. Figure 4.7 illustrates the biochemical reactions, the logic symbol, and the logic truth table for an intercellular gate that implements the IMPLIES logic function. In the absence of the input mRNA and its corresponding repressor, RNAp, binds to the promoter and transcribes the output gene, yielding a high output. As with the inverter, if only the input repressor is present, it binds to the promoter and prevents transcription, yielding a low output. Finally, if both the repressor and the inducer are present, the

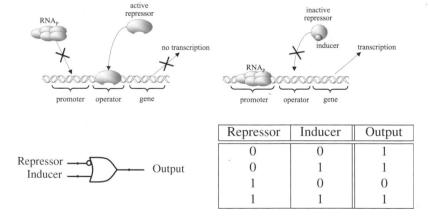

Repressor	Inducer	Output
0	0	1
0	1	1
1	0	0
1	1	1

Figure 4.7 A genetic gate for the IMPLIES logic function using repressors and inducers. Shown here are the two states when the repressor protein is present, the logic symbol for the gate, and the logic truth table.

inducer binds to the repressor and changes the conformation of the repressor. The conformation change prevents the repressor from binding to the operator and allows RNAp to transcribe the gene, yielding a high output.

The IMPLIES gate has the same three biochemical stages as the inverter: translation, cooperative binding, and transcription. The inducer concentration levels affect the cooperative binding stage C', which now has two inputs. Let υ_A denote the inducer concentration level, let ϕ_A denote the concentration level of the input mRNA, and let ϕ_Z denote the concentration level of the output mRNA. Here, the repression activity ρ_A resulting from the cooperative binding stage, $\rho_A = C'(\phi_A, \upsilon_A)$, depends nonlinearly on the level of the repressor and on the level of the inducer. The binding affinities of the active repressor to the operator, and of the inducer molecule to the repressor, determine the shape of C'. The transfer function $\mathcal{I}$ of the IMPLIES logic gate is the mapping:

$$\psi_Z = \mathcal{I}(\psi_A) = \mathcal{T} \circ C' \left[\upsilon_A, \mathcal{L}(\psi_A)\right].$$

The two gate inputs are not interchangeable. The input and output repressors can be connected to any other circuit component, but the inducer input is an intercellular signal and is specifically coupled to the input repressor. As with the different inverter signals, before building a circuit the designer should experimentally check for unintended interactions between a specific inducer, other repressors, and other inducers. Any set of noninterfering repressor/inducer pairs can then serve as a library of potential signals for constructing intercellular circuits.

The AND Gate

The AND gate allows cells to detect incoming messages from other cells. Figure 4.8 illustrates the biochemical reactions, the logic symbol, and the truth table for an intercellular gate that implements the AND logic function. Here, RNAp normally has a low affinity for the promoter, and basal transcription is correspondingly small. Therefore, in the absence of the activator and inducer inputs, the output is low. If the activator is present but the inducer is not present, the activator has a low affinity for the promoter and does not bind to it. In this case, the output is still low. Finally, if the inducer and the activator are both present, the inducer binds to the activator. The newly formed bond changes the conformation of the activator and allows the activator/inducer complex to bind to the operator. In turn, the activator/inducer complex helps recruit RNAp to the promoter and initiate transcription, yielding a high output. Often, a dimeric form of the activator is necessary for complex formation.

Similar to the biochemical stages of the NOT and IMPLIES gates, the AND gate stages include translation, cooperative binding, and transcription. The first stage, translation, is similar in all three cases. For the AND gate, the following cooperative binding stage C'' maps the activator protein ϕ_A and inducer υ_A

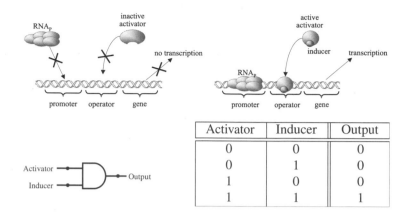

Activator	Inducer	Output
0	0	0
0	1	0
1	0	0
1	1	1

Figure 4.8 A genetic gate for the AND logic function using activators and inducers. Shown here are two of the states when the activator is present, the logic symbol for the gate, and the logic truth table.

inputs to an *activation* level, ω_A, rather than repression. Similar to the two-input cooperative binding stage for the IMPLIES gate, the activation level $\omega_A = C''(\phi_A, \upsilon_A)$ depends nonlinearly on the concentration of activator and inducer. The binding affinities of the inducer molecule to the activator and of the activator/inducer complex to the operator determine the shape of C''. Last, the transcription stage maps the activation level, ω_A, to output mRNA, ψ_Z, in a positive relation. The transfer function $\mathcal{I}$ of the AND logic gate is the mapping:

$$\psi_Z = \mathcal{I}(\psi_A) = \mathcal{T} \circ C'' \left[\upsilon_A, \mathcal{L}(\psi_A) \right].$$

As with the IMPLIES gate, the two inputs are not interchangeable, and the inducer must always be coupled with the corresponding activator. Again, the interference between inducers and other circuit elements should be checked experimentally before circuit design. Beyond the challenge of finding non-interfering activator/inducer pairs, any multicellular system design must also address the spatial issues of intercellular communications.

The SEND Gate

To initiate communications to receiver cells that have the AND gate, sender cells synthesize inducer molecules using a SEND gate. The input to the SEND gate is mRNA coding for a special catalytic enzyme, with a concentration of ϕ_A. The enzymatic reaction of the gate, $\mathcal{E}$, produces inducer with an intracellular concentration υ_A, where the inducer also diffuses into the medium. The steady-state mapping, $\upsilon_A = \mathcal{E}(\phi_A)$, is a positive relation with an asymptotic boundary defined by the availability of the appropriate metabolic precursors in the cell.

The intracellular level of the inducer, υ_A, is also affected by the surrounding extracellular concentration of the inducer.

The next section describes biocircuit design and analysis with BioSPICE.

BIOCIRCUIT DESIGN AND ANALYSIS

Biocircuit design poses several important challanges, some that are in common with electrical circuit design, and some that are unique. As with electrical circuits, the behavior of biocircuits depends on the characteristics of the component gates. It is therefore important to first characterize the behavior of the gates by measuring their device physics before attempting to design circuits of significant complexity. Based on the characteristics of individual gates, one can predict the behavior of complex circuits built from these components. Here, we use BioSPICE to develop and simulate a model of biochemical inversion and then demonstrate how to predict the behavior of biocircuits based on this model. We also show that the correct behavior of these circuits is highly dependent on using gates with matching characteristics.

Initial attempts at constructing gates often yield devices with unacceptable behavior. For example, a particular device may have insufficient signal restoration capabilities or inadequate noise margins. In designing electrical circuits, one uses process engineering to modify the characteristics of the devices (e.g., gain or trigger levels) until they obtain the desired behavior. In this section, we introduce genetic process engineering, the analogous mechanism for biocircuit design. We demonstrate how BioSPICE facilitates genetic process engineering by predicting the new behavior of devices that results from genetic modification to specific stages in biochemical inversion. The analysis and insights gained from these BioSPICE predictions guide the engineer in genetically modifying existing components to achieve the appropriate behavior for building reliable circuits of significant complexity.

In contrast to electrical circuit design where identical components are separated spatially, each component in a biocircuit shares the same physical space but relies on different biochemical reactions. The complexity of biocircuit design is exacerbated by the fact that the components typically have widely varying kinetic characteristics. These gates are built from DNA-binding proteins, ribosome-binding sites, and protein-binding sites with inherently different kinetic characteristics. Therefore, a critical element in biocircuit design is analyzing and optimizing the behavior of each new gate included in the cellular gate library. In this section, we describe the conditions necessary for matching gates to achieve correct digital behavior. This analysis motivates specific genetic modifications to achieve gate-device physics that match with other gates for correct design and construction of complex circuitry.

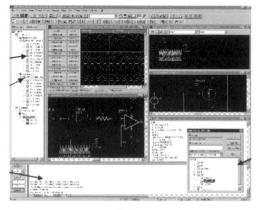

Spice

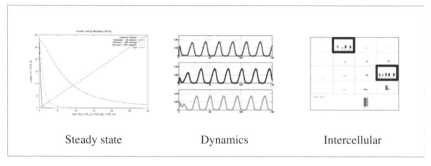

Steady state   Dynamics   Intercellular

BioSPICE

Figure 4.9 Tools for circuit design: Spice and BioSPICE.

In discussing biocircuit design, this section often highlights the role of BioSPICE. With electrical circuits, tools such as SPICE [10] help designers manage the complexity of their substrate and achieve reliable systems. The BioSPICE prototype tool aims to help biocircuit designers in a similar manner. It is an integrated collection of software tools that features analysis of the steady state, dynamics, and intercellular behavior of genetic logic circuits (Figure 4.9). BioSPICE enables the designer to analyze and simulate modifications to the behavioral characteristics of cellular gates. It can thus provide valuable insights toward reliable system design, such as matching gate input/output threshold levels.

A Biochemical Model of an Inverter

To implement digital-logic computation, an inverter combines several natural gene regulatory mechanisms. These mechanisms include transcriptional

Table 4.1 Biochemical reactions that model an inverter.

$$\text{mRNA}_A + \text{rRNA} \xrightarrow[\text{translate}]{k_{xlate}} \text{mRNA}_A + \text{rRNA} + A \qquad (1)$$

$$\text{mRNA}_A \xrightarrow[\text{decay}]{k_{dec(mrna)}} \qquad (2)$$

$$A + A \xrightarrow[\text{dimerization}]{k_{dim(a)}} A_2 \qquad (3) \qquad\qquad A \xrightarrow[\text{decay}]{k_{dec(a)}} \qquad (5)$$

$$A_2 \xrightarrow[\text{single}]{k_{sngl(a)}} A + A \qquad (4) \qquad\qquad A_2 \xrightarrow[\text{decay}]{k_{dec(a2)}} \qquad (6)$$

$$P_Z + A_2 \xrightarrow[\text{repress1}]{k_{rprs(a2)}} P_Z A_2 \qquad (7) \qquad\qquad P_Z A_4 \xrightarrow[\text{decay}]{k_{dec(ga4)}} P_Z A_2 \qquad (10)$$

$$P_Z A_2 \xrightarrow[\text{dissociation}]{k_{dis(a2)}} P_Z + A_2 \qquad (8) \qquad\qquad P_Z A_2 + A_2 \xrightarrow[\text{repress2}]{k_{rprs(a4)}} P_Z A_4 \qquad (11)$$

$$P_Z A_2 \xrightarrow[\text{decay}]{k_{dec(ga2)}} P_Z \qquad (9) \qquad\qquad P_Z A_4 \xrightarrow[\text{dissociation}]{k_{dis(a4)}} P_Z A_2 + A_2 \qquad (12)$$

$$P_Z + \text{RNA}_p \xrightarrow[\text{transcribe}]{k_{xscribe}} P_Z + \text{RNA}_p + \text{mRNA}_Z \qquad (13)$$

$$\text{mRNA}_Z \xrightarrow[\text{decay}]{k_{dec(mrna)}} \qquad (14)$$

mRNA_A is the input and mRNA_Z the output.

control, translation of mRNA, repression through cooperative binding, and degradation of proteins and mRNA transcripts.

Table 4.1 presents one possible chemical model of the reactions involved in biochemical inversion. In particular, this model incorporates characteristics from the bacteriophage λcI repressor operating on the P(R) promoter and the $O_R 1$ and $O_R 2$ operators. The mRNA_A molecule represents the input signal, and the mRNA_Z molecule represents the output signal. Ribosomal RNA (rRNA) translates mRNA_A into the input protein repressor A, and A_2 denotes the dimeric form of A. P_Z denotes the concentration of the active form of the promoter for Z. A promoter is active only when its associated operator is unbound by a repressor. $P_Z A_2$ and $P_Z A_4$ represent the repressed (i.e., inactive) forms of the promoter, where either one or two dimers is bound to the promoter, respectively. RNA polymerase (RNAp) initiates transcription from the active form of the promoter, P_Z, into mRNA_Z, the gene transcript.[2] This gene transcript typically codes for other signals (e.g., protein repressors or activators), or for structural and enzymatic proteins that perform certain cellular tasks.

---

[2] The simulations in this section assume that the concentrations of RNA_p and rRNA are fixed. Chapter 7 discusses how to measure the effect of fluctuations in these concentrations, as well as other factors, on the inverter's behavior. Once these effects have been quantified, robust gates can be designed.

The model includes the components described in the second section of this chapter in the following reactions: translation of the input protein from the input mRNA (reactions 8–9), input protein dimerization and decay (reactions 10–13), cooperative binding of the input protein (reactions 14–19), transcription (reaction 20), and degradation of the output mRNA (reaction 21).

To simulate an inverter, BioSPICE automatically converts the chemical equations of this model to the system of ordinary differential equations in Table 4.2. The system of differential equations includes an entry for each of the molecular species involved in inversion: the input mRNA, $mRNA_A$; the input protein monomer, A; the dimeric form of the input repressor protein, A_2; the unbound promoter, P_Z; the promoter bound with one repressor protein dimer, $P_Z A_2$; the promoter bound with two repressor protein dimers, $P_Z A_4$; and the output mRNA, $mRNA_Z$. Each differential equation describes the time-domain behavior of a particular molecular species based on all the equations in the biochemical model that include that particular molecule. For example, the "$k_{dim(a)} \cdot A^2$" term of differential Equation 17 is derived from model Equation 3, while the "$-k_{sngl(a)} \cdot A_2$" term is derived from model Equation 4. Note that the

Table 4.2 Ordinary differential equations used to simulate a single inverter.

$$d(mRNA_A) = drive_{mRNA_A}(t) - k_{dec(mrna)} \cdot mRNA_A \tag{15}$$

$$d(A) = 2 \cdot k_{sngl(a)} \cdot A_2 - k_{dec(a)} \cdot A + k_{xlate} \cdot rRNA \cdot mRNA_A$$
$$- 2 \cdot k_{dim(a)} \cdot A^2 \tag{16}$$

$$d(A_2) = k_{dim(a)} \cdot A^2 - k_{sngl(a)} \cdot A_2 - k_{dec(a2)} \cdot A_2$$
$$- k_{rprs(a2)} \cdot P_Z \cdot A_2 + k_{dis(a2)} \cdot P_Z A_2$$
$$- k_{rprs(a4)} \cdot P_Z A_2 \cdot A_2 + k_{dis(a4)} \cdot P_Z A_4 \tag{17}$$

$$d(P_Z) = k_{dis(a2)} \cdot P_Z A_2 - k_{rprs(a2)} \cdot P_Z \cdot A_2 + k_{dec(ga2)} \cdot P_Z A_2 \tag{18}$$

$$d(P_Z A_2) = k_{rprs(a2)} \cdot P_Z \cdot A_2 - k_{dis(a2)} \cdot P_Z A_2 - k_{rprs(a4)} \cdot P_Z A_2 \cdot A_2$$
$$+ k_{dec(ga4)} \cdot P_Z A_4 - k_{dec(ga2)} \cdot P_Z A_2 + k_{dis(a4)} \cdot P_Z A_4 \tag{19}$$

$$d(P_Z A_4) = k_{rprs(a4)} \cdot P_Z A_2 \cdot A_2 - k_{dis(a4)} \cdot P_Z A_4 - k_{dec(ga4)} \cdot P_Z A_4 \tag{20}$$

$$d(mRNA_Z) = k_{xscribe} \cdot P_Z \cdot RNA_p - k_{dec(mRNA)} \cdot mRNA_Z \tag{21}$$

These equations are derived from the biochemical inversion model in Table 4.1.

"drive$_{\text{mRNA}_A}(t)$" term in differential Equation 15 allows the user to specify an external drive to vary the input levels of the inverter over time. The user can observe the dynamic behavior of the circuit by directing BioSPICE to solve the system of differential equations using MATLAB's stiff differential equations solver, ode15s [15]. This solver is a variable order solver based on the numerical differentiation formulas that optionally uses backward differentiation formulas (Gear's method).

Figure 4.10 shows a BioSPICE simulation of the dynamic behavior of the inverter circuit with the above chemical reactions in response to an external stimulus. The kinetic constants (Table 4.3) used in this simulation were obtained from the literature describing the phage λ promoter P_R and repressor (cI) mechanism [4, 13]. The graphs show the concentrations of the molecules involved in the inversion, with blue curves representing mRNA, red curves representing proteins, and green curves representing genes. The top graph is for the input mRNA$_A$, followed by graphs for the input protein repressor and its dimeric form, followed by graphs for the active and inactive forms of the gene, and finally the graph for the output mRNA$_Z$.

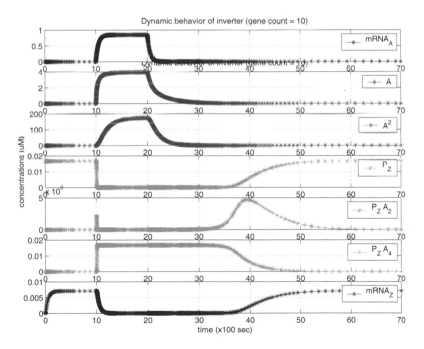

Figure 4.10 The dynamic behavior of the inverter. The graphs show a time-series of the molecular concentrations involved in inversion in response to a stimulus of input mRNA.

Table 4.3 Kinetic constants used in BioSPICE simulations.

$k_{dim(a)}$	8.333	$k_{rprs(a2)}$	66.67	$k_{dec(a)}$	0.5775	$k_{dec(ga2)}$	0.2887
$k_{sngl(a)}$	0.1667	$k_{dis(a2)}$	0.2	$k_{dec(a2)}$	0.5775	$k_{dec(ga4)}$	0.2887
$k_{dim(z)}$	8.333	$k_{rprs(a4)}$	333.3	$k_{dec(z)}$	0.5775	$k_{dec(mrna)}$	2.0
$k_{sngl(z)}$	0.1667	$k_{dis(a4)}$	0.25	$k_{dec(z2)}$	0.5775	$k_{xscribe}$	0.0001
						k_{xlate}	0.03

The units for the first-order reactions are 100 sec^{-1}, and the units for the second order reactions are $\mu M^{-1} \cdot 100 \text{ sec}^{-1}$.

The reactions proceed as follows: At first, no input mRNA or input protein repressor are present. As a result, P_Z is active and RNAp transcribes the gene into the output $mRNA_Z$. The level of $mRNA_Z$ increases until it stabilizes when the gene expression and decay reactions reach a balance. At this stage, the input signal is low while the output signal is high.

Then, an externally imposed drive increases the input $mRNA_A$, which rRNA translates into the input repressor protein A. The protein begins to form dimers, and these dimers bind to the promoter's free operators. The system quickly reaches a state where each promoter is essentially completely bound by two dimers. The almost total inactivation of the promoters occurs at a fairly low concentration of the dimer A_2 and indicates the strong repression efficiency of the cI repressor that is used for this simulation. As a result of the promoter inactivation, transcription stops, and the output $mRNA_Z$ decays to zero. At the end of this stage, the input signal is high while the output signal is low.

Finally, the external drive of $mRNA_A$ stops, resulting in the decay of $mRNA_A$, A, and A_2. Slowly, the repressor dimers dissociate from the P_Z operators, and the level of the active promoter P_Z rises back to the original level. This allows RNAp to resume transcription of P_Z, and the level of the output $mRNA_Z$ rises again. At this stage, the input signal reverts to low, while the output signal reverts to high.

The simulation shows that the gate switching time (measured in minutes for this mechanism) is governed by the rate of recovery of P_Z. The limiting rate is therefore the dissociation of bound repressor dimer A_2 from P_Z. Figure 4.11 shows simulation results of how an engineered reduction in the repressor binding coefficient improves the gate delay. Through simulations such as this, BioSPICE offers insights into biocircuit design and ultimately motivates laboratory experiments. For example, chapter 7 describes experimental results of reducing the binding efficiency of cI to the λ promoter by mutating base pairs in the O_R1 region.

Simulations of Proof of Concept Circuits

This section describes BioSPICE simulation results of an RS latch and a ring oscillator, two simple but interesting logic circuits built from the inverter model

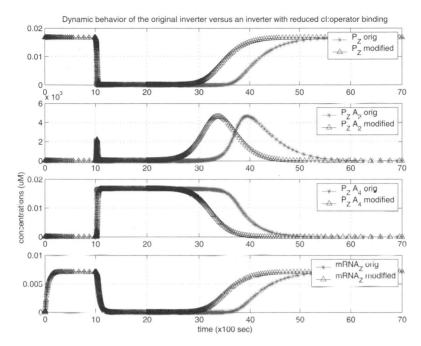

Figure 4.11 Improving gate delay: the effect of reducing the repressor binding efficiency by a factor of 100. For the promoter and output mRNA, the graphs compare the molecular concentrations of the original gate and of the modified gate. Overall, the modification reduces the gate delay in switching from low to high output.

discussed above. The simulations provide a first indication of the feasibility of using the proposed chemical reactions to implement logic circuits. Furthermore, the analysis and simulations of modifications to the kinetic coefficients provide valuable insights toward reliable system design.

Storage: Analysis of an RS Latch

The RS latch is a good initial test circuit for the biochemical inversion model and should operate correctly even if its constituent parts are not perfectly matched. It persistently maintains a data bit that can be toggled on and off. The RS latch consists of two cross-coupled NAND gates, with inputs $\overline{S}$ and $\overline{R}$ for setting and resetting the complementary output values A and B (Figure 4.12). The inverters with inputs $\overline{R}$ and B and common output A constitute one of the NAND gates, while the inverters with inputs $\overline{S}$ and A and common output B constitute the other NAND gate. The inputs $\overline{S}$ and $\overline{R}$ are normally high, and are set to low to toggle the latch.

Figure 4.13 shows the correct simulated dynamic behavior of this RS latch in response to relatively short and long input pulses. Initially, both inputs $\overline{S}$ and $\overline{R}$ are high, and the outputs A and B compete for dominance. In this simulation,

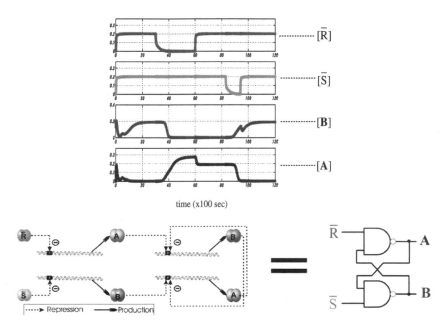

time (x100 sec)

Figure 4.12 Dynamic behavior and circuit diagrams of the RS latch. The inputs $\overline{S}$ and $\overline{R}$ are normally high and are set to low to toggle the state of the outputs A and B. The simulations shows that the gate operates correctly in response to relatively long and short input pulses.

B becomes high while A becomes low. In a physical implementation of this circuit, factors such as the relative repression efficiency, original concentration level, and stochastic noise determine which signal initially becomes high.

After the initial output values settle into a steady state, an external stimulus reduces the level of the input $\overline{R}$ to toggle the value stored by the latch. This relatively long pulse results in expression of the output A and a subsequent decay of the output B. When $\overline{R}$ regains its original high level, the circuit still maintains a high level for A and a low level for B. Notice that the expression of A from two genes during the toggle phase results in a level of A that is higher than than the level of A during the steady state following the toggle. However, the circuit functions correctly because the higher analog value of A does not exceed the range defined to be a digital 1. Because A is a repressor, once A reaches a saturating repression level, any additional increases in concentration do not affect the output of the gate.

Finally, a short external stimulus reduces the level of $\overline{S}$ to toggle the RS latch back to the original state. In this case, $\overline{S}$ regains its high level before B builds up to its own steady-state high level. The level of B drops for a short period,

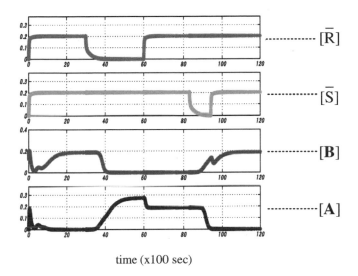

time (x100 sec)

Figure 4.13 Dynamic behavior of the RS latch. The top two curves are the $\overline{\text{reset}}$ and $\overline{\text{set}}$ inputs, respectively. The bottom two curves are the complementary outputs. The initial behavior shows the system settling into a steady state.

but then B rises back up to the appropriate high level. Therefore, as expected, both long and short pulses effectively set and reset the latch.

Connections: Analysis of a Ring Oscillator

A ring oscillator is another useful test circuit, especially because the correct behavior of the circuit is highly sensitive to the device physics of its components. The oscillator consists of three inverters connected in a series loop without any external drive (Figure 4.14). The simulation results in Figure 4.15 depict the expected oscillation in signal concentrations, as well as a phase shift between the values (values shown are protein concentrations, not mRNA). Note, however, that oscillation occurs close to the low end of the signal values. This results from the skewed transfer curve that describes the steady-state characteristics of the inverters. Basically, for each inverter in the circuit, a low level of the input

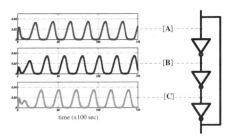

Figure 4.14 Dynamic behavior of a ring oscillator. The three curves are the outputs of the three inverters. Note the 120° phase shift between successive stages.

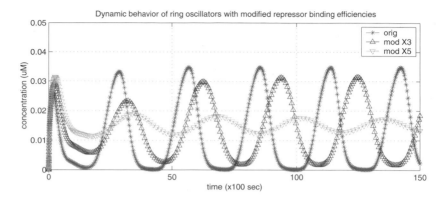

Figure 4.15 Dynamic behavior of ring oscillators with modified repressor binding coefficients.

repressor is sufficient to inactivate the corresponding promoter. Once the input of an inverter reaches that threshold, the inverter's output will begin to decay.

The circuit oscillates correctly when the gates are perfectly matched, but incorrect behavior may result from coupling mismatched components. Figure 4.16 shows the effect of mismatched inverters on the dynamic behavior of the ring oscillator. The inverters have different binding coefficients and transcription rates. Specifically, the values of the kinetic constants $k_{rprs(2)}$ and $k_{rprs(4)}$ for protein repressor B are now a third of the original cI values, and the value of $k_{xscribe}$ for C's promoter is now twice the original. As a result, the output of the inverter with the strongest transcription rate settles into high,

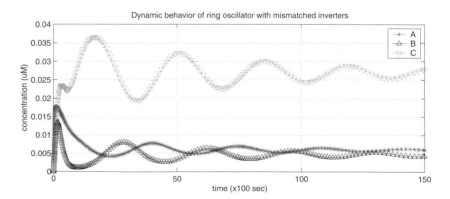

Figure 4.16 A time-series simulation illustrating the incorrect behavior of a ring oscillator with mismatched inverters. The second inverter's repressor binding coefficient is three times lower than the original, while the third inverter's transcription rate is twice as strong as the original.

while the other two outputs settle into low. Clearly, correct behavior of the circuit is highly sensitive to the device physics of its components. Therefore, an integrated approach using BioSPICE and laboratory experiments is crucial for the success of biocircuit design.

Steady State: Design and Analysis

As shown above, the correct dynamic behavior of biocircuits is highly dependent on the steady-state characteristics of the component gates. The complexity of biocircuit design is exacerbated by the fact that each of the components uses different protein and DNA-binding sites. These typically have widely varying kinetic characteristics. Therefore, a critical element in biocircuit design is analyzing the steady-state behavior of the cellular gates. The analysis motivates specific genetic modifications to achieve gate device physics that match with other gates for correct design and construction of complex circuitry. This section describes BioSPICE simulations to compute the transfer curve of an inverter, the steady-state conditions necessary for matching gates, BioSPICE analysis of genetic mutations to optimize the inverter's steady-state characteristics, and prediction of the behavior of circuits using transfer functions of individual inverters.

Steady-State Behavior

The transfer curve of an inverter, useful for analyzing the steady-state behavior, maps an input level ϕ_A to an output level ϕ_Z. Figure 4.17 shows the transfer curve of an inverter with the kinetic rates from Table 4.3, as computed by BioSPICE. To compute the transfer curve of a given circuit, BioSPICE performs a set of simulations, where each simulation has a different steady rate of input signal synthesis. For each synthesis rate, BioSPICE records the level of the corresponding output signal if the system settles into a steady state. In this case, the definition of a steady state is some number of simulation time steps where all the state variables do not fluctuate by more than a small threshold. Each simulation that settles into a steady state contributes a point to the approximation of the transfer curve. The inset in Figure 4.17 illustrates the transfer curve of two such inverters connected in series. The skewed curve shows that an inverter based on the λcI repressor and P(R) promoter is very sensitive to low concentrations of the input protein. Thus, for a high output signal, even small increases in the synthesis rate of the input may alter the output signal to low.

Matching Thresholds

An inverter's transfer curve describes the behavior of a single gate, but the information is also useful for connecting multiple gates into operational logic circuits. Transfer functions suitable for implementing digital-logic circuits gates must have low and high ranges such that signals in the low range map strictly into the high range, and vice versa. The shape of the curve, represented by its gain

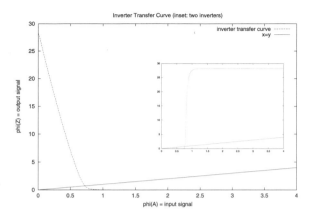

Figure 4.17 The simulated transfer curve of the $\lambda \, cI/P(R)$ inverter, with the transfer curve of two such inverters connected in series shown in the inset. Both graphs plot ϕ_A versus ϕ_Z.

(i.e., slope), determines how well the gate reduces noise from input to output. For electronic digital circuits, the low and high signal ranges are typically the same for all gates because the circuit is composed of transistors with identical threshold voltages, spatially arranged. However, in biochemical digital circuits, the gate components (e.g., proteins and promoters) have different characteristics depending on their reaction kinetics. Therefore, the designer of biological digital circuits must take explicit steps to ensure that the signal ranges for coupled gates are matched appropriately, as described below.

Before discussing the rules for biochemical gate coupling, we introduce a variation of the transfer function, the *transfer band*, that captures systematic fluctuations in signal levels. It is especially important to consider fluctuations in biological settings. Experiments in chapter 7 report signal fluctuations by a factor of 10 for cells with the same genetic circuit grown under the same conditions. The transfer band captures these fluctuatations with a region enclosed by a pair of transfer functions, as shown in Figure 4.18. $\mathcal{I}^{\min}$ is the function that maps an input to the minimum corresponding observed output, and $\mathcal{I}^{\max}$ is the function that maps an input to the maximum corresponding observed output.

Let I_{il} and I_{ih} be the input thresholds. Then, the low and high gate matching requirement from above is:

[in low]	$\langle 0, I_{il} \rangle$	$\xrightarrow{\text{into}}$	$\langle \mathcal{I}^{\min}(I_{il}), \mathcal{I}^{\max}(0) \rangle$	[out high]
[in high]	$\langle I_{ih}, \infty \rangle$	$\xrightarrow{\text{into}}$	$\langle 0, \mathcal{I}^{\max}(I_{ih}) \rangle$	[out low]

Consider the case of two inverters, I and J, with J's output coupled to I's input. Then, the coupling is correct if and only if:

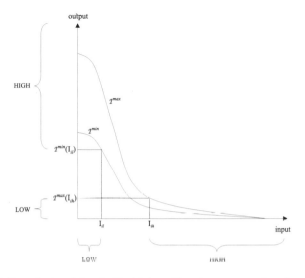

Figure 4.18 Transfer band thresholds: high and low input ranges for a hypothetical inverter. The transfer band, capturing systematic fluctuations in signals, is defined by the two curves.

$$\langle \mathcal{J}^{\min}(J_{il}), \mathcal{J}^{\max}(0) \rangle \quad \subset \quad \langle I_{ih}, \infty \rangle$$

$$\langle 0, \mathcal{J}^{\max}(J_{ih}) \rangle \quad \subset \quad \langle 0, I_{il} \rangle.$$

Then, assuming monotonic functions, the following conditions are necessary and sufficient for correct coupling:

$$\mathcal{J}^{\min}(J_{il}) \quad > \quad I_{ih}$$

$$\mathcal{J}^{\max}(J_{ih}) \quad < \quad I_{il}.$$

Genetic Process Engineering

The first step in developing biocircuits is designing, building, and characterizing several inverters. It is likely that these inverters will not match correctly according to the definitions above. Fortunately, there are biochemical techniques to adjust inverters to obtain the correct behavior for use in complex circuits (Figure 4.19). These are described below.

MODIFYING THE REPRESSOR/OPERATOR BINDING AFFINITY   Certain basepair mutations at an operator site alter the affinity of the repressor to the operator. As illustrated in Figure 4.19a, weakening the repressor affinity shifts $\mathcal{C}$, the cooperative binding stage mapping the input signal, ϕ_A, to the repression activity, ρ_A, outward to the right. For a given level of the input repressor protein, there is now less repression activity than before the mutation. The mutations

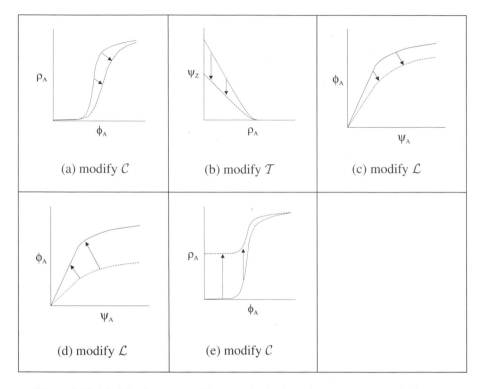

Figure 4.19 Modifications to specific stages in the inversion process: (a) reducing the repressor/operator binding affinity; (b) reducing the strength of the promoter; (c) reducing the strength of the ribosome binding site; (d) increasing the cistron count; (e) adding autorepression.

result in different behaviors for each repressor/operator pair. Figure 4.20 shows BioSPICE simulations where hypothetical reductions in the repression affinity of cI to O_R1 shift the overall transfer function of the cI/P(R) inverter outward. In these simulations, the original values of $k_{rprs(a2)}$ and $k_{rprs(a4)}$ were scaled down by factors of 10, 100, and 1000. Chapter 7 reports experimental transfer functions results where the affinity of cI to the λ promoter is modified by mutating base pairs on OR_1. The experimental results demonstrate the desired shift outward.

MODIFYING THE STRENGTH OF THE PROMOTER   Certain basepair mutations at the promoter site alter the affinity of the RNA polymerase to the promoter. DNA sequence determinants of promoter strengths have been studied extensively [2, 14, 17]. As illustrated by Figure 4.19b, weakening a promoter shifts T, the transcription-stage mapping between ρ_A and ψ_Z, inward. Any given level of the active promoter (the complement of ρ_A) now results in less transcription and

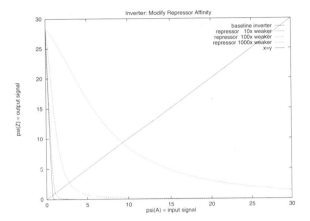

Figure 4.20 Reducing the binding affinity of the repressor to the operator shifts the transfer function of an inverter outward.

therefore less mRNA output (ψ_Z). BioSPICE simulations (Figure 4.21) show the overall inward shifts of the transfer curve of the cI/P(R) inverter that result from weakening the P(R) promoter. In these simulations, the original value of $k_{xscribe}$ was scaled down by factors of 2, 5, and 10.

MODIFYING THE STRENGTH OF THE RIBOSOMAL BINDING SITE (RBS)   Modifications to the RBS alter the affinity of rRNA to the mRNA. Figure 4.19c illustrates the shift in $\mathcal{L}$, the translation stage mapping between ψ_A and ϕ_A, resulting from a reduction in rRNA affinity. For any given level of the input mRNA, ψ_A, there

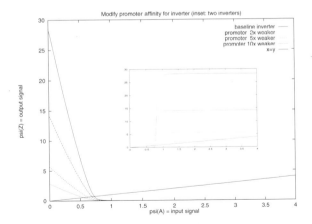

Figure 4.21 Reducing the binding affinity of the RNA polymerase shifts the transfer functions of the cI/P(R) inverter inward. Inset shows the effects on the transfer functions of two inverters in series. The diagonal lines correspond to input equals output.

is now less translational activity and therefore less input protein, ϕ_A. Chapter 7 reports on experimental transfer curve results with different ribosome binding sites for *cI*. Weaker RBSs caused an outward shift to the overall transfer curve of the original *cI*/P(R) inverter.

INCREASING THE CISTRON COUNT   The cistron count is increased by adding copies of the coding sequence for the output protein downstream of the promoter. As illustrated in Figure 4.19d, the increase in cistron count shifts $\mathcal{L}$, the translation stage mapping between ψ_A and ϕ_A, upward. For any given level of the mRNA input, ψ_A, there is now additional input protein repressor ϕ_A. The curve values are scaled by a linear factor in the initial range of the translation before reaching metabolic constraints of the cell's translational machinery. BioSPICEs simulations (Figure 4.22) demonstrate the overall outward shift of the transfer curve of the *cI*/P(R) inverter as the cistron count increases from one to five.

ALTERING THE DEGRADATION RATE OF A PROTEIN   The degradation rate of proteins can be normally achieved by changing a few amino acid residues on the C terminus [1, 11, 12]. For example, decreasing the half-life of the input protein A causes $\mathcal{L}$, the translation stage mapping between ψ_A and ϕ_A, to shift downward. Because of the increase in the protein degradation rate, for any given level of the input mRNA ψ_A, there is now less repressor protein, ϕ_A, present. Another possibility for increasing the degradation rate is to choose bacterial strains that have high concentrations of effective proteases [16].

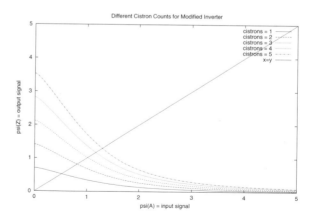

Figure 4.22 The effects of increasing the cistron count per gene (i.e., the number of structural genes per operator region). In this case, the original inverter mechanism has been modified to 100× less binding affinity of the repressor and 40× less binding affinity of the promoter.

ADDING AUTOREPRESSION   An operator that binds the output protein may be added to the promoter/operator region to accomplish autorepression. The simulated transfer functions shown above are not balanced because they are more sensitive to fluctuations in low input signals versus fluctuations in high input signals. Autorepression reduces the maximum output protein synthesis rate and therefore reduces some of the asymmetry in the signal sensitivities. Figure 4.19e illustrates the effect of adding autorepression on C, the cooperative binding stage of inversion. For low levels of the input protein repressor, the promoter is active and transcribes the output protein. This output protein binds to the operator of its own promoter, and therefore there is always some concentration of inactive/bound promoter, ρ_A. High levels of input repressor increase ρ_A until saturation.

Functional Composition of Transfer Functions

In biocircuit design, the engineer creates a genetic logic circuit using a small set of basic gates and a database of biochemical reaction kinetic rates. The kinetics database contains transfer-curve measurements obtained using the mechanism described in chapter 7. Given transfer-curve measurements, the engineer predicts the behavior of complex circuits through the functional composition of behavioral data describing only the basic components. For example, figures 4.17 and 4.21 show the predicted steady-state behavior of circuits with two inverters based on the transfer curves of the constituent inverters.

Without autorepression, the transfer function of an inverter is determined by the input mRNA, ϕ_A, RBS, input protein A, operator, promoter, and output mRNA, ϕ_Z, but not by the output protein Z. This means that the relation $\phi_Z = \mathcal{I}_1(\phi_A)$ does not depend on Z. Therefore, gate couplings are independent of the output protein. To predict the transfer function of two gates in a series, $\phi_Y = \mathcal{I}_2(\phi_Z)$ connected to $\phi_Z = \mathcal{I}_1(\phi_A)$, the database only needs to contain $* = \mathcal{I}_2(\phi_Z)$ and $* = \mathcal{I}_1(\phi_A)$, where $*$ denotes any protein. If the inversion uses autorepression, then the relation $\phi_Z = \mathcal{I}_3(\phi_A)$ also depends on the characteristics of Z. To compute the transfer function of a gates coupling $\phi_Y = \mathcal{I}_4(\phi_Z)$ connected to $\phi_Z = \mathcal{I}_3(\phi_A)$, the database must specifically include the transfer functions of the input/output protein pairs Y/Z and Z/A and their associated mRNA.

Using the transfer-curve measurements, the efficacy of a particular transfer function in implementing the digital abstraction is evaluated in terms of factors such as gain and noise margins. The transfer function of inverters in a series is the functional composition of their respective transfer functions. A series of inverters is then also evaluated in terms of gain and noise margins. Because the transfer functions are different for each inverter, the gates must be matched. In addition, the matching process must also evaluate the gain and noise margins resulting from gate couplings and only select proteins that achieve satisfactory levels of these factors.

Chapter 7 describes measurements and genetic modifications of *in vivo* logic gates to obtain components with the desired behavior for constructing complex and reliable circuits.

CONCLUSIONS

This chapter lays the foundation of an engineering discipline for obtaining complex, predictable, and reliable cell behaviors by embedding biochemical logic circuits and programmed intercellular communications into cells. To accomplish this goal, this chapter provides a well-characterized cellular gate library, a biocircuit design methodology, and software design tools. The cellular gate library includes biochemical gates that implement the NOT, IMPLIES, and AND logic functions in *E. coli* cells.

We introduced a biocircuit design methodology that comprises a mechanism for measuring the device physics of gates and criteria for evaluating, modifying, and matching gates based on their steady-state behavior. By using the abstraction of logic circuits, complex and reliable behavior is synthesized from reliable, well-characterized components with matching input/output characteristics.

In addition to the above contributions, we have developed BioSPICE, a prototype software tool for biocircuit design. BioSPICE simulates *in vivo* logic circuits using ordinary differential equations that model biochemical rate equations. The kinetics for the rate constants were derived from the literature and yield simulation results that predict the behavior of engineered biocircuits. BioSPICE simulations of modified rate constants illustrate the effects on the static and dynamic behavior of the circuits and serve as motivation for genetically modifying components in laboratory experiments.

References

[1] J. U. Bowie and R. T. Sauer. Identification of c-terminal extensions that protect proteins from intracellular proteolysis. *J. Biol. Chem.*, 264(13):7596–7602, 1989.

[2] D. E. Draper. Translational initiation. In Frederick C. Neidhardt, editor, *Escherichia coli and Salmonella*, 2nd ed., pages 902–908. ASM Press, Washington, DC, 1992.

[3] H. A. Greisman and C. O. Pabo. A general strategy for selecting high-affinity zinc finger proteins for diverse dna target sites. *Science*, 275:657–661, 1997.

[4] R. W. Hendrix. *Lambda II*. Cold Spring Harbor Laboratory Press, Cold Spring Harbor, NY, 1983.

[5] A. Hjelmfelt, E. D. Weinberger, and J. Ross. Chemical implementation of neural networks and Turing machines. *Proc. Natl. Acad. Sci. USA*, 88:10983–10987, 1991.

[6] T. F. Knight Jr. and G. J. Sussman. Cellular gate technology. In *Proceedings of UMC98: First International Conference on Unconventional Models of Computation*, pages 257–272. Auckland, NZ, January 1998. Springer-Verlag, Singapore.

[7] H. H. McAdams and A. Arkin. Stochastic mechanisms in gene expression. *Proc. Natl. Acad. Sci. USA*, 94:814–819, 1997.

[8] H. H. McAdams and A. Arkin. Simulation of prokaryotic genetic circuits. *Annu. Rev. Biophys. Biomol. Struc.*, 27:199–224, 1998.

[9] H. H. McAdams and L. Shapiro. Circuit simulation of genetic networks. *Science*, 269:650–656, 1995.

[10] L. W. Nagel. SPICE2: A computer program to simulate semiconductor circuits. Technical Report ERL Memo. No. UCB/ERL M75/520, University of California at Berkeley, May 1975.

[11] A. A. Pakula and R. T. Sauer. Genetic analysis of protein stability and function. *Annu. Rev. Genet.*, 23:289–310, 1989.

[12] D. A. Parsell, K. R. Silber, and R. T. Sauer. Carboxy-terminal determinants of intracellular protein degradation. *Genes Devel.*, 4:277–286, 1990.

[13] M. Ptashne. *A Genetic Switch: Phage Lambda and Higher Organisms*, 2nd ed. Cell Press and Blackwell Scientific Publications, Cambridge, 1986.

[14] M. T. Record Jr., W. S. Reznikoff, M. L. Craig, K. L. McQuade, and P. J. Schlax. *Escherichia coli* RNA polymerase ($e\sigma^{70}$), promoters, and the kinetics of the steps of transcription initiation. In Frederick C. Neidhardt, editor, *Escherichia Coli and Salmonella*, 2nd ed., pages 792–821. ASM Press, Washington, DC, 1992.

[15] L. F. Shampine and M. W. Reichelt. The MATLAB ODE suite. *SIAM J. Sci. Computing*, 18:1–22, 1997.

[16] J. R. Swartz. *Escherichia coli and Salmonella: Cellular and Molecular Biology*, vol. 2, 2nd ed., pages 1693–1711. ASM Press, Washington DC, 1996.

[17] P. H. von Hippel, T. D. Yager, and S. C. Gill. *Quantitative Aspects of the Transcription Cycle in Escherichia coli*, pages 179–201. Cold Spring Harbor Laboratory Press, Cold Spring Harbor, NY, 1992.

[18] S. A. Ward and R. H. Halstead Jr. *Computation Structures*. MIT Press, Cambridge, 1990.

5

The Device Science of Whole Cells as Components in Microscale and Nanoscale Systems

Michael L. Simpson, Gary S. Sayler,
James T. Fleming, John Sanseverino,
and Chris D. Cox

Intact whole cells may be the ultimate functional molecular-scale machines, and our ability to manipulate the genetic mechanisms that control these functions is relatively advanced when compared to our ability to control the synthesis and direct the assembly of man-made materials into systems of comparable complexity and functional density. Although engineered whole cells deployed in biosensor systems provide one of the practical successes of molecular-scale devices, these devices explore only a small portion of the full functionality of the cells. Individual or self-organized groups of cells exhibit extremely complex functionality that includes sensing, communication, navigation, cooperation, and even fabrication of synthetic nanoscopic materials. Adding this functionality to engineered systems provides motivation for deploying whole cells as components in microscale and nanoscale devices. In this chapter we focus on the device science of whole cell components in a way analogous to the device physics of semiconductor components. We consider engineering the information transport within and between cells, communication between cells and synthetic devices, the integration of cells into nanostructured and microstructured substrates to form highly functional systems, and modeling and simulation of information processing in cells.

INTRODUCTION

Even a casual examination of the information processing density of prokaryotic cells produces an appreciation for the advanced state of the cell's capabilities. A bacterial cell such as *Escherichia coli* (2 μm^2 cross-sectional area) with a 4.6 million basepair chromosome has the equivalent of a 9.2-megabit memory. This memory codes for as many as 4300 different polypeptides under the inducible control of several hundred different promoters. These polypeptides perform metabolic and regulatory functions that process the energy and information, respectively, made available to the cell. This complexity of functionality allows the cell to interact with, influence, and, to some degree, control its environment.

Compare this to the silicon semiconductor situation as described in the International Technology Roadmap for Semiconductors (ITRS) [54]. ITRS predicts that by the year 2014, memory density will reach 24.5 Gbits/cm^2, and logic transistor density will reach 664 M/cm^2. Assuming four transistors per logic function, 2 μm^2 of silicon could contain a 490-bit memory or approximately three simple logic gates. Even this level of functionality depends on an unsure path of technology development that will require breakthroughs in lithography, materials, processing, defect detection, and many other challenging areas. We can confidently say that silicon technology will not approach bacterial-scale integration within the foreseeable future.

Aside from this functional density, microorganisms have other attributes desirable for engineered synthetic systems. Cells are relatively rugged components that subsist in extreme environments such as deep sea thermal vents, subzero arctic seawaters, hypersaline solutions, water saturated with organic solvents, contaminated soils, and industrial wastes. Prokaryotic cells are relatively easy to manipulate genetically, and they have a diverse set of gene regulation systems that allow their inclusion in the types of hybrid systems considered here. Furthermore, these cells can easily be incorporated into a 3D structure (i.e., 3D integration) instead of the 2D structure of integrated circuits (ICs). And cells self-replicate and self-assemble into groups (e.g., biofilms), making them easy to manufacture with no requirement of lithography, mask alignment, or other technologically challenging processing steps to produce highly functional components.

COMPLEXITY OF FUNCTIONALITY: WILD-TYPE ORGANISMS

Before proceeding down the path of engineered cellular components, it is instructive to consider the complex functionality exhibited in nature, both by individual cells and by the more complex behavior of interconnected groups of cells. Here we present three examples.

Magnetotaxis

Individual cells are capable of performing extremely complex tasks as illustrated by magnetotaxis, a type of directed motility. Many microorganisms in aqueous environments are motile in an active search for nutrients. Some bacteria are capable of a tactic response to various chemical (chemotaxis) or physical (e.g., phototaxis) stimuli. Bacterial magnetotactic behavior was discovered more than 25 years ago [12]. The examination of motile aquatic bacteria showed that a magnetic field comparable to the geomagnetic field is enough to control the direction of travel [38].

Magnetotactic bacteria fabricate nanoparticles of magnetite (Fe_3O_4 [7, 39, 66, 68, 107]) or greigite (Fe_3S_4 [6, 37, 50, 65]), enclose them in membranes (magnetosomes [3]), and assemble these into linear arrays. This linear array of magnetosome nanoparticles constitutes a permanent magnetic dipole fixed within the bacterium [38] that is large enough to force orientation, and therefore travel, along geomagnetic field lines.

Formation of Symbiotic Relationships

Sensing, information processing, and actuation by microorganisms might involve communication with higher organisms. Such symbiotic relationships are usually formed through a process in which a host becomes colonized by specific microorganisms from the surrounding environment. The mutual benefits derived from these associations by host and symbionts are well documented. However, only recently have the mechanisms by which the partners make contact and establish a symbiosis been described. This process often requires complex sensing, communication, information processing, and actuation from the microorganisms.

For example, consider the inoculation of the Hawaiian squid, *Euprymna scolopes*, with the luminescent bacteria *Vibrio fischeri*. *E. scolopes* has a light-emitting organ that only functions when colonized by *V. fischeri* within its mantle cavity. During the inoculation process, the squid ventilates 1.3 μl of seawater, which on average contains only one *V. fischeri* cell, through the mantle cavity every 0.5 sec [76]. If inoculation occurred only by random processes, during this 0.5-sec period the *V. fischeri* would have to find their way to one of only six 10-μm pores that lead into the light organ [76]. However, experiments performed by Nyholm et al. [76] provided evidence that two-way communication between *V. fischeri* and *E. scolopes* mediates accumulation of *V. fischeri* and promotes the colonization of the light organ. To enhance the symbiotic relationship, *V. fischeri* exhibit a loss of flagellation, reduction in cell size, decrease in growth rate, and enhancement of luminescence soon after inoculation [88]. These events are apparently catalyzed by *V. fischeri* sensing the light organ environment and processing this information through changes in gene expression [111].

Biofilm Formation

The sensing, information processing, and actuation behavior of groups of microorganisms might be even more complex than that found in the formation of a symbiotic relationship. It is now understood that many bacteria use molecular signaling to initiate group behavior that is significantly more complex than single organism behavior. These systems employ small, diffusible molecules to accomplish cell-to-cell communication that leads to group behavior such as the formation of biofilms or group bioluminescence. The organization of microbial communities in biofilms is generally considered a predominant life form of many prokaryotic and lower eukaryotic populations. Rather than an amorphous assemblage of individual populations, biofilms exhibit structured formation, assembly, and morphology and represent a rudimentary model of prokaryotic development [24], in some ways analogous to tissues.

Recent developments in biofilm research have been extensively reviewed [22, 26, 27]. What has become clear over the past decade is that there is a well-recognized structural dynamic leading to mature biofilms, and there is significant extracellular communication and control among members of biofilm communities. The group nature of biofilms provides protection for individual cells that allows survival in a hostile environment. Members of the biofilm enjoy some protection from phage, biocides [16, 20] or potent antibiotics [75]. Furthermore, the members of the biofilm community cooperate to promote survival of the colony. Each cell in the community provides specialized functionality in a complex community that has primitive homeostasis, a primitive circulatory system, and cooperative metabolic activity [21].

The preceding three examples illustrate the significant abilities of cells to sense, communicate, navigate, cooperate, and even to fabricate synthetic nanoscopic materials. With these capabilities cells have the ability to interact with, influence, and, to some degree, control their environment in a way much coveted by designers of artificial systems. Having briefly considered this functionality in natural systems, we now turn to a consideration of cells as components in microscale and nanoscale systems.

CELLS AS COMPONENTS IN ENGINEERED SYSTEMS

There is a significant body of experience in the creation of sensing, information processing, and actuating devices and systems from the combination of components constructed from a variety of materials. Examples would include silicon transistors, aluminum interconnects, silicon dioxide insulators, and carbon resistors combined to form an electronic circuit. Fundamental properties of these components determine both how they can be arranged within an information pathway to produce a desired response to a given stimulus and how they can be physically arranged on a common substrate. As components in these systems, cells reside in this environment, and the same component issues must be addressed. These issues fall into four main categories:

Information processing: what are the information processing pathways, and how can these be manipulated to produce a desired result?

Communications: what are the modes of communication (interconnectivity, input/output), and how can they be manipulated to allow for the realization of complex functionality through high interconnectivity?

Fabrication and integration: how can the components be physically arranged, packaged, and their functionality maintained or even enhanced within micro- and nanostructured substrates?

Design, modeling, and simulation: how can the devices be modeled in a way that allows the simulation required for complex system design?

Apart from the scientific pursuit of elucidating cell function, the answers to these questions, from the point of view of the device designer, define the device science of whole cells as components in microscale and nanoscale systems.

Cells as Bulk and Discrete Components

The cellular components in whole-cell devices can be used as bulk components that rely on the average activity of groups of cells. Such devices usually perform a chemical or biological sensing function where the bulk reaction of cells to a particular stimulus is detected by a synthetic system. Cell-based sensing systems often couple a molecular-sensing element to a reporter gene through gene fusion, thereby allowing the use of regulatory proteins and promoter sequences of chromosomal or plasmid DNA as the molecular-sensing elements [18]. The specificity of these elements produce system selectivity and analytical sensitivity, while the reporter gene product, in combination with the synthetic transducer, determine system sensitivity and detection limits.

An alternate approach would be to use individual or small groups of cells as discrete components in engineered systems. This would include larger collections of cells, such as biofilms or tissues, where individual or small groups of cells perform information processing that lead to group behavior through interconnection of these discrete elements. The information transport path shown in Figure 5.1 illustrates the contrast between cells as bulk and discrete components. The use of whole cells as bulk or discrete components greatly affects the functional density (i.e., amount of information processing per unit area), the noise performance, input/output strategies, packaging, and through these, the ultimate application space of particular cell-based systems. To this point the dominant use of whole cells in microscale systems has been as bulk components, the major exception being tissue- or biofilm-based devices. Therefore, many of the technological challenges such as packaging and integrating cells into these microscale systems have been addressed in the context of bulk whole-cell devices.

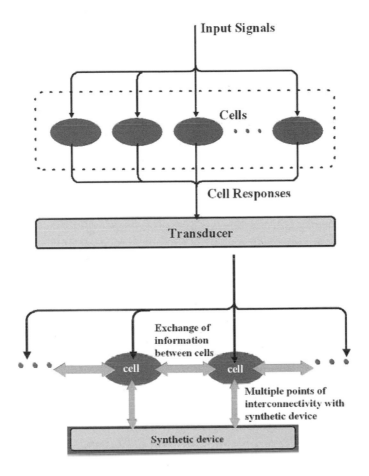

Figure 5.1 Contrast between the use of cells as bulk or discrete components. (a) Cells deployed as a bulk component process information in isolation from each other and their responses are averaged. (b) Cells used as discrete components have multiple points of interconnectivity with a synthetic device and may include cell-to-cell connectivity.

However, attaining for engineered devices the complex functionality described above for wild-type organisms requires the use of cells as discrete components with a molecular-scale interface between the natural and synthetic elements of the system. In the text that follows, we will present the characterization of whole cells as components in engineered systems gained from their deployment as bulk elements. However, we must travel toward the use of cells as discrete components to fulfill the full promise of this technology, and it is in this direction that we ultimately look in this chapter.

INFORMATION PROCESSING IN CELLS

Information Processing with Natural Genetic Circuits

Gene expression profiling, which provides powerful analyses of transcriptional responses to cellular perturbation and interconnectivity among gene regulation in transcription, represents one important use of the information processing functionality of cells. Most often this is performed by screening isolated mRNA using cDNA microarrays. Uses of this technology have included analyses uncovering the consequences of individual mutations [34], understanding cellular physiology under various conditions [55], examining disease states [33], and measuring responses to pharmaceuticals [67] and crop protection products [56]. However, there are important limitations to these microarray methods, particularly for the types of systems considered here. First, the mRNA isolation step would be difficult to implement in a small, low-power, and automated manner. Furthermore, artifacts may arise during the RNA isolation [104] or from cross-hybridization [85]. And finally, as a destructive measurement, only a single snapshot in time of the expression profile can be obtained.

An important alternative to genome-wide expression profiling uses reporter gene fusions. Transcriptional fusions have been widely used [60, 92] because they can be straightforwardly produced [17], identified, and assayed [58, 112]. Reporter genes are described in more detail in the next section. Recently, the identification and mapping of a large number of random *Escherichia coli* DNA fragments fused to luminescent reporter genes was reported [110]. The result was a genome-wide, genome-registered collection of *Escherichia coli* bioluminescent reporter gene fusions that mirrors the transcriptional wiring diagram of *E. coli*. Information enters the cells through interaction with the genetic regulatory machinery, is processed by the existing genetic circuits, and discrete outputs are generated by reporter gene expression. While presenting many of the same advantages, this collection of intact whole-cell components avoids many of the disadvantages of the DNA microarray technology. In particular, no RNA extraction or similar manipulation is required, and rather than a snapshot, a time history of gene expression (or at least protein activity) is generated.

Silicon Mimetic Approach

While the genome-wide, genome-registered *E. coli* collection described above provides an important tool for using the information processing capabilities of whole cells, only the natural genetic circuits and wiring of these cells is used. We describe here what we have coined the "silicon mimetic" approach, where genetic circuit functionality is engineered to perform functions analogous to those found in man-made systems.

Silicon semiconductor technology is the gold standard of man-made information processing devices and systems. In such devices, information is

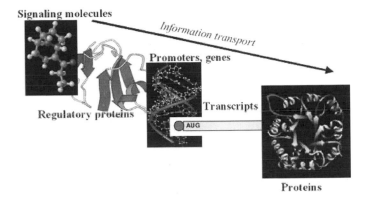

Figure 5.2 Information transport through an individual genetic regulatory circuit. The representation of the information is initially coded in the concentration of small molecules. During processing the format of this representation changes several times with the result ultimately represented by the concentration or activity of a protein.

represented by voltages and currents and is communicated through the controlled transport of electrons. In contrast, information in cellular systems is represented by molecular concentrations and is communicated through the transport of these molecules and control of molecular production by genetic expression. As information is transported through cells, the manner in which it is represented changes several times (Figure 5.2), in contrast to silicon systems where the information always remains coded within a collection of electrons.

Transistors are the fundamental building block for silicon information processing devices. These three-terminal components control current flow between two terminals using the voltage on the third terminal (Figure 5.3a). Hiratsuka and co-workers [51, 52] have proposed the realization of transistor-like circuits and interconnections using enzyme-catalyzed reactions and diffusion of the products.

Hiratsuka et al.'s proposed fundamental device is roughly analogous to a bipolar junction transistor with current flow represented by the enzymatic conversion of a substrate into a product as controlled by effectors (Figure 5.3b). These investigators propose to create complex functionality by coupling multiple enzyme transistors to form a network of biochemical reactions defined by the molecular selectivity of enzyme transistors. Information would be coded into molecular agents and then discriminated by the selectivity of enzyme transistors. In this plan, information is transported from point to point by molecular diffusion. Because different molecules represent individual quanta of information, there is no need for insulation between interconnect paths. This is an increasingly difficult and ultimately limiting problem for silicon systems at higher

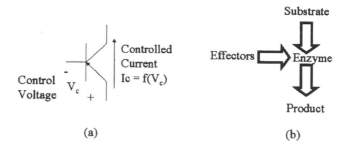

Figure 5.3 Transistor action of (a) a semiconductor device and (b) an enzymatic device. The semiconductor transistor controls the flow of current between two terminals using a voltage on the third terminal, while the enzyme transistor controls the conversion of a substrate into a product using the concentration of effectors required for the conversion.

levels of complexity where 10 levels of lithographically defined interconnects are called for by the year 2014 [54].

A scheme similar to the enzyme networks described above could be envisioned for cellular systems. Figure 5.4 shows the biochemical reactions responsible for light production in the prokaryotic *lux* system. This reaction implements an enzyme transistor with myristyl aldehyde as the substrate, reduced flavin mononucleotide (FMNH$_2$) and O$_2$ as co-effectors, and light, flavin mononucleotide (FMN), and myristic acid as products. The entire reaction can be thought of as several interconnected enzyme transistors at work. Additionally, because the production of the enzymes that catalyze these reactions is controlled by expression of the *lux* genes, genetic regulatory functions involving inducers, regulatory proteins, and promoters are part of the circuit.

The transistor-like devices described above are useful for realizing biosensors [94] and may provide some of the conceptual framework for whole-cell circuits with more utility. Engineered Boolean logic functions realized within the genetic circuits of cells are more extensively treated elsewhere in this volume. We briefly review some of these efforts here.

Logic gates are devices that compute Boolean algebraic functions. The inputs and outputs are logic levels (i.e., true/false or one/zero), and the outputs are derived from the inputs through the application of a set of simple rules. For example, consider the AND, OR, and XOR gates shown in Figure 5.5. The output of an AND gate is true only if both of its inputs are true. Likewise, the output of an OR gate is true if either of its inputs are true, whereas an XOR gate has a true output if one, but not both, of its inputs are true. Although each of these gates provide very modest computational power, as demonstrated by silicon integrated circuit technology, the interconnection of these devices can lead to exceptional functionality.

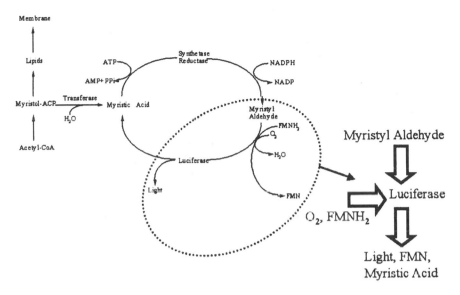

Figure 5.4 The biochemical reaction responsible for the production of bioluminescence in the bacterial Lux system. The indicated portion of this reaction realizes an enzyme transistor with myristyl aldehyde as the substrate; light, flavin mononucleotide (FMN), and myristic acid as the products, and O_2 and $FMNH_2$ as effectors.

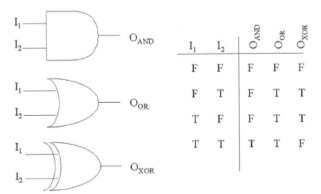

Figure 5.5 AND, OR, and XOR logic gates. The outputs of these devices are determined by applying simple rules to the input signals as shown in the truth tables.

Some recent work has centered on constructing gene transcription modules to create biochemical devices that can be combined to create logic circuits 51, 94, 114, 115]. The transcriptional unit consists of a promoter that is inducible or controllable and a gene. The gene product can be a regulator or an enzyme. The inputs are molecular signals that control gene expression, transcription, or translation, whereas the output of the logic gate can be the enzyme itself or the activity of the enzyme.

As a simple example, consider the implementation of a transcriptional OR gate. There are at least two strategies for implementing the OR gate: (1) use two promoters that have identical gene transcriptional effects but respond to two different inducers; or (2) use a single promoter that responds in a similar manner to two different inducers. The latter strategy has been implemented using a *tod-lux* fusion in *Pseudomonas putida* TVA8 [94]. This operon is similarly induced by both trichloroethylene (TCE) and toluene [91]. The result is the production of bioluminescence if either inducer is present. As Figure 5.6 shows, this construct implements the logical OR function.

The AND, OR, and XOR functions described above are functionally complete as they can be combined to implement any combinatorial logic function. However, these components cannot be used to implement sequential circuits that require memory of past logic states and clock signals for synchronization. Recently, Gardner et al. [42] demonstrated a two-state genetic latch in *E. coli*

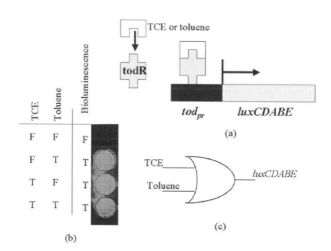

Figure 5.6 A simple OR gate constructed using a *tod-lux* fusion. (a) The todR regulatory protein reacts similarly to either trichloroethane (TCE) or toluene, thereby upregulating the expression of the *lux CDABE* genes and producing bioluminescence. (b) A truth table for this simple genetic logic gate where the presence (1) or absence (0) of bioluminescence represents the logic state of the output. (c) Logic symbol representation of this genetic circuit.

that implements a one-bit memory using two repressible promoters arranged in a mutually inhibitory network. A significant challenge still remains to develop genetic circuits that make more efficient use of the memory capacity of the cell's DNA. Likewise, Elowitz and Leibler [35] implemented a genetic clock circuit in *E. coli* with a typical period of hours. Such genetic constructs might eventually be useful for clocking simple sequential circuits in whole cells. However, for now this clock displays noise and cell-to-cell variations that make its use in functional logic circuits problematic.

Noise and Random Processes in Cellular Information-Processing Systems

Information processing through genetic circuits is subject to stochastic processes leading to variability in response to apparently identical stimuli. This is not a concern for cells deployed as bulk components, as the response is averaged over a large population of cells. However, in complex systems where the response of individual or small collections of cells becomes important, the stochastic nature of gene regulation must be considered and incorporated into models and simulations to enable design.

Biochemical reactions function as components in highly organized and regulated networks, not as isolated processes. Cellular signals control the transcriptional events that lead eventually to translation of these genes into proteins at the first level within these networks. Some of these proteins will be regulatory and will affect the transcription of additional genes. All of these molecules have a dynamic half-life that determines the time of their effectiveness within the network. The response interval between these cellular events is determined by the time required for concentrations of each component in the network to increase or decrease to their effective concentration ranges. In other words, a finite period of time is required for an effector to reach an operational concentration and, likewise, a certain period of time is required for that effector to decay below the operational concentration [69].

Experimental evidence of variability both in molecular concentrations and response time interval suggests a degree of random fluctuation at the macro level characteristic of the processes that occur in all chemical reactions at the micro level. The result at the macro level is that the outcomes of genetic networks are not entirely deterministic [2]. A corollary to this is that population–averaged variations in gene expression are due to changes in the frequency of full gene induction in individual cells rather than to uniform variations in gene expression across the entire population. The term *stochastic* has been applied to these fluctuations implying statistical variability [69]. These stochastic processes produce noise in the genetic circuits discussed here. The result could be nothing worse than small predictable variations from average responses in linear circuits. However, in more complex, nonlinear circuits involving genetic pathway bifurcation with considerably different final conditions, this could lead to probabilistic outcomes considerably different from simple Boolean logic rules described above.

Solutions to the stochastic formulation of coupled reactions can be computed using the Monte Carlo procedure described by Gillespie [43]. This algorithm calculates a stochastic description of the temporal behavior of the coupled reactions by calculating the probabilistic outcome of each discrete chemical event and the resulting changes in the number of each molecular species. By accumulating the results for all reactions over time, the statistics of the inherent fluctuations in the reaction products can be estimated [69]. Deterministic information-processing circuits engineered into these cells must deal with the stochastic nature of these nanoscale chemical systems through reduction of noise by feedback [8] or through the development of information-processing algorithms that make use of probabilistic rather than deterministic outcomes.

INTERCONNECTIVITY AND INPUT/OUTPUT

Engineering information transport within individual cells only partially answers the challenges associated with deploying cells as components in engineered devices. To construct systems of even moderate complexity requires the interconnection of these information pathways within cells and cellular communities and between cells and the synthetic portions of these systems. Figure 5.7 illustrates the information transport pathways through the type of hybrid systems considered here. Three communication pathways and interfaces must be considered (we will use the generic term *chip* to refer to all of the possible synthetic components that may be interfaced to the cells): (1) cell to chip; (2) cell to cell; and (3) chip to cell.

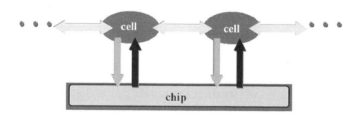

Figure 5.7 Communication pathways in a hybrid whole-cell/synthetic device system. Three separate communication pathways are seen: (1) cell to chip, (2) cell to cell, and (3) chip to cell.

Cell-to-Chip Communication

Cell-to-chip communication may be accomplished through the use of reporter genes, electrochemical means, or through the recording of action potentials. Reporter gene expression produces a signal that readily couples to the chip. To be useful in the devices considered here, the reporter gene product or activity must be readily detected by the chip and fused to the appropriate engineered information pathway(s) of the cell. Furthermore, while a discrete output (e.g., expressed or not expressed) would have some uses, an output that is continuously variable with expression level allows more flexibility in design. Here we focus on the use of reporter genes in cell-to-chip communication. Comprehensive reviews of reporter gene applications may be found elsewhere [25, 74, 122].

There are seven unique reporter protein candidates for cell-to-chip communication: (1) chloramphenicol acetyltransferase (CAT), (2) β-galactosidase, (3) aequorin, (4) firefly luciferase, (5) green fluorescent protein (GFP), (6) uroporphyrinogen III methyltransferase (UMT), and (7) bacterial luciferase (Lux). CAT, β-galactosidase, aequorin, and firefly luciferase require the addition of an exogenous substrate, which is a significant complication for cell-to-chip communication as it is considered here. This leaves GFP, UMT, and Lux as the most viable means of cell-to-chip communication. Each of these is briefly described below.

Lux

Bioluminescent organisms include species of bacteria, algae, dinoflagellates, fungi, jellyfish, clams, fish, insects, shrimp, and squid. Luminescent bacteria are classified into three genera: (1) *Xenorhabdus*, (2) *Photobacterium*, and (3) *Vibrio* [73]. The relative ease of transferring cDNA coding for Lux proteins into prokaryotic and eukaryotic organisms has resulted in their common use as reporter genes. Bacterial luciferase catalyzes the oxidation of $FMNH_2$ and a long-chain fatty aldehyde to FMN and the corresponding fatty acid in the presence of molecular oxygen (Figure 5.4). This reaction produces blue-green light with a maximum intensity at 490 nm and quantum efficiency between 0.05 and 0.15 [72]. Fatty aldehydes of chains ranging from 7 to 16 carbons long may serve as substrates for the reaction [72].

The α and β structural subunits of bacterial luciferase are encoded by the *luxA* and *luxB* genes, and, with the use of an exogenous substrate, the expression of these two genes results in bioluminescence. However, three additional genes in the *lux* operon, C, D, and E, code for proteins required for synthesis and recycling of the fatty aldehyde [11, 120]. Thus, the expression of all five *lux* genes results in bioluminescence without the addition of an exogenous substrate, and for the systems considered here, the entire *luxCDABE* cassette is required.

Green Fluorescent Protein

The green fluorescent protein has been isolated and cloned from the jellyfish *Aequorea victoria*. Due to its internal covalently bound imidazolinone chromophore, GFP is autofluorescent, and the addition of cofactors or exogenous substrates is not required to produce light [74]. The maxima of the GFP excitation spectrum resides at 395 nm with a minor peak at 475 nm, while the emission maxima occurs at 509 nm with a small shoulder at 540 nm [116]. Although, the fluorescence quantum yield of GFP (0.72–0.85) is comparable to that of fluorescein (0.91), its lower molar absorptivity leads to an approximately 1 order of magnitude lower intensity than fluorescein [108].

Advantages of GFP include high stability at biological pH and lack of endogenous homologues in most target organisms [25, 74]. However, it has recently been reported that GFP is toxic to some mammalian cell types, and this cytotoxicity may be linked to apoptosis [63]. Still, the unique characteristics of GFP have led to its successful application in gene expression studies, transformed cell identification, and to use as a reporter gene in whole-cell biosensors. Unfortunately, the requirement of an excitation source complicates the use of this reporter in the types of hybrid systems considered. However, GFP may be used in conjunction with luciferase, shifting the emission spectrum from blue-green bioluminescence (490 nm peak) to green fluorescence (509 nm peak) [82].

The stability and spectral properties of GFP may be modified through structural alterations of the wild-type protein [25, 57]. Notably, several mutants with altered excitation and emission spectra have been developed, including blue-, red-, cyan- and yellow-shifted variants. This spectral diversity provides potentially important flexibility for cell-to-chip communication by allowing parallel transmission channels.

Uroporphyrinogen (Urogen) III Methyltransferase

UMT has been identified and purified from several different organisms and exists in two forms [90]. The first form is encoded by the *cobA* genes in *Bacillus megaterium* [86], *Methanobacterium ivanovii* [14], *Propionibacterium freudenreichii* [90], and *Pseudomonas denitrificans* [13], and the second form is encoded by the *cysG* gene in *E. coli* [113] and *Salmonella typhimurium* [44]. UMT catalyzes the *S*-adenosyl-L-methionine (SAM)-dependent addition of two methyl groups to the substrate, urogen III. This produces dihydrosirohydrochlorin (precorrin-2), which can be oxidized to a fluorescent product, sirohydrochlorin, or once again through UMT action accept the addition of a third methyl group yielding a second fluorescent product, trimethylpyrrocorphin. Both products emit an orange-red to red (590–770 nm) fluorescence in response to UV light excitation at 300 nm [87]. The urogen III substrate can be found in all organisms [87], allowing UMT to function as a reporter gene without requiring the addition of a substrate or other cofactors other than the

UV excitation. As in the GFP case, this requirement of an excitation source is a disadvantage for the types of systems considered here.

Although the intensity of the fluorescent signal produced by the UMT system is comparable to that of GFP [118], the red fluorescent properties may provide a greater signal-to-noise ratio because autofluorescence and light scattering of endogenous materials are lower in this spectral region [118]. Furthermore, the quantum efficiency of chip-based photodetectors typically reaches a maximum in this spectral region while having almost no sensitivity to the UV excitation [93]. For these reasons, UMT may prove to be a useful reporter protein for cell-to-chip communication.

Ionic/Electronic Communication

Techniques for performing noninvasive recordings from cultured cells (e.g., cardiomyocytes, neuronal networks) using microelectrode arrays have been developed during the last two decades [80]. Action potentials generated by the cells are capacitively coupled to electrodes arranged in a 2D array on a substrate surface [48]. Recordings from neuronal networks cultured on these microelectrode arrays exhibit highly complex spatially and temporally distributed signal patterns that are highly sensitive to their environment. Recent work has focused on using such systems for chemical and environmental sensing and for pharmaceutical screening [32, 97]. It should be noted that for networks of cells, in addition to cell-to-chip coupling, cell-to-cell communication is involved in generating the complex signals found at the electrodes.

Electrochemical Communication

Microelectrophysiological electrochemical measurements (i.e., electroanalytical detection techniques) performed at ultramicroelectrodes (<50 μm) are emerging as a promising avenue for examining and monitoring chemical dynamics at the single-cell level [53]. These electrodes exhibit fast (millisecond) response times, high mass sensitivity (zeptomole), small size, large linear dynamic range (up to 4 orders of magnitude), and selectivity. Furthermore, molecules of interest can be followed without the need for derivitization as is necessary in fluorescence microscopy. The microelectrodes reported to date for single-cell analysis typically consist of a linear carbon nanotube bundle surrounded by an insulating layer of glass. Macroscopic carbon bundles are placed into glass capillary tubes, which are then pulled down to microscale dimensions (0.5–10 μm). Microscopic inspection allows the cleaving of the pulled capillary in the vicinity of the entrained carbon bundle, providing a conductive carbon tip, surrounded by an insulating sheath of glass. This microelectrode can be placed in close proximity to an extracellular or intracellular region of interest and used to analyze the local microenvironment for a large variety of electroactive species. For easily oxidizable substances, such as catecholamines [117], indoleamines [31], oxygen [61], and doxorubicin [64], the native carbon electrode surface is sufficient for electrochemical analysis, and these substances have been readily detected in or near the surface of single cells. In fact, these electrodes, at present, are the only available technique to measure the exocytotic release of neurotransmitters from single cells [1]. An overview of the application of electrochemistry with ultra-microelectrodes in neuronal microenvironments has been compiled by Clark et al. [19]. In chapter 8 we present progress toward the realization of molecular-scale electrochemical probes suitable for integration into whole-cell microscale and nanoscale systems.

Cell-to-Cell Communication

In addition to the coupling of information between cells using the action potentials described above, communication of signals between whole-cell components can be accomplished through the naturally occurring chemical signals used by microbial cells to coordinate population-level processes. Communication via small molecular intermediates has been demonstrated to regulate several diverse processes in bacteria including development, conjugation, pathogenesis, antibiotic production, symbiosis, competence, and bioluminescence. Evidence for cell-to-cell communication in bacteria was first obtained for the process of *quorum sensing* (QS). QS is a cell-cell communication network that enables bacterial populations to collectively regulate their behavior based on cell density, serving as a way for bacteria to coordinate their metabolic efforts for specific functions such as infection onset, antibiotic production, or biofilm formation [30, 47, 121]. This intercellular exchange relies on self-generated, low-molecular-weight, diffusible signal molecules called *autoinducers* (AIs). AI molecules released by a single, free-living bacterium do not accumulate to high enough concentrations to be detected. However, when a sufficient bacterial population density is reached, AI concentrations achieve a threshold level that allows individual bacteria to coordinately activate or repress specific gene expression.

QS was first identified in the marine bacterium *Vibrio fischeri*, where it controls expression of bioluminescence [99]. The *V. fischeri* QS network consists of AI molecules called *N*-acylhomoserine lactones (AHLs). AHLs are synthesized from precursors by a synthase protein, LuxI and, upon reaching a critical threshold, interact with a transcriptional activating LuxR protein to induce expression of genes responsible for bioluminescence. Bacteria use QS to control a variety of cell-density–dependent functions, including virulence [119], iron regulation [15], 100], and biofilm formation [27, 29]. One of the best-characterized QS model systems available is that of *Pseudomonas aeruginosa*, an opportunistic pathogen [89] that can serve as a model organism for pathogenic processes. The *P. aeruginosa* QS system derives from two pathways, designated *las* and *rhl*, which function in a hierarchical nature to regulate more than 40 genes involved in the formation of virulence determinants and secondary metabolites involved in biofilm formation.

The primary considerations for molecular interconnection schemes are the number of different molecules available and the level of cross-talk between the different molecular species. Research over the last 10 years has demonstrated that AHL signaling is widespread among Gram-negative bacteria, having been reported in more that 29 species of the α, β, and γ subdivisions of the proteobacteria [103]. Twenty-three individual LuxI homologs have been reported, and the regulatory components have been shown to be conserved at the genetic level. All of the AHLs characterized to date vary primarily in the length and hydrophobicity of their acyl side chains [41]. Experimental evidence suggests a high degree of specificity between the AHL and its cognate transcriptional activator. The V. fisheri LuxR showed no activity in the presence of the P. aeruginosa AHL, and, similarly, the P. aeruginosa LuxR homolog, LasR, showed no activity in the presence of the V. fisheri AHL [46].

However, more recently, evidence for cross-species communication has begun to appear in the literature. In one study in which several species of bacteria were tested for production of extracellular AHL-like activities using V. harveyi sensor mutants, V. cholerae and V. parahaemolyticus were shown to exhibit such activities [5]. Also, V. harveyi was shown to have two independent systems, one highly species specific and another species nonspecific [5]. Using the V. harveyi sensor mutant, it was shown that several strains of E. coli and S. typhimurium produced AHL-like molecules [101]. It appears that the species-specific system monitors the environment for other V. harveyi, whereas the nonspecific system monitors for other species of bacteria [4].

The variety of independent AHL-like signaling processes might permit massive parallel interconnection between whole-cell information-processing devices. A species-specific AHL could be produced that would diffuse throughout a population of cells and yet only react with a single target cell or some subset of the total population. Similarly, a species nonspecific AHL could be used to coordinate processes in an entire population of cells. This ability to both broadcast and target communication provides a significant tool for the realization of highly complex whole-cell devices.

CHIP-TO-CELL COMMUNICATION

The communication of information from chip to cell may be the most challenging of the three major information pathways presented here. However, this information pathway is essential for the complete integration of cells as components in engineered systems, and it possibly represents the most far-reaching step beyond that accomplished in whole-cell biosensor systems. Here we consider communication that may proceed in two ways: (1) information transport that interacts with the functioning of cellular processes other than gene regulation (e.g., ionic transport); and (2) information transport through the manipulation of gene expression.

Electronic/Ionic Communication

Direct electronic communication with ionic transport in neurons has been demonstrated [40]. Signals were supplied to neurons through a capacitor structure composed of a p-doped silicon electrode with a thin-film oxide between the electrode and the cell. A voltage pulse applied to the silicon electrode elicited an action potential in the neuron. The simplest type of stimulation was a positive voltage step applied to the doped silicon, inducing a negative charge at the portion of the cellular membrane in contact with the silicon capacitor and a positive charge on the portion of the membrane most distant from the capacitor. The intracellular response was an exponentially decaying voltage step [98]. With properly designed structures, this intracellular voltage pulse had a sufficient amplitude and duration to trigger an action potential [98]. Silicon structures have been constructed that both stimulate and record the action potential from an individual neuron [98]. The same investigators have constructed a device that transported a signal from the silicon substrate into an action potential, which elicited an action potential in a neighboring neuron that was subsequently recorded by a transistor on the silicon substrate [124]. This was a significant advance as this structure demonstrated all three major communication pathways in a single hybrid device.

Information Transport into Genetic Regulatory Circuits

ELECTRICALLY INDUCIBLE PROMOTERS   While chemical means may be used to communicate with genetic regulatory circuits, physical mechanisms would be advantageous for whole-cell systems interfaced to physical systems. Optical excitation may be used to regulate both bacterial and eukaryotic photosynthesis genes or constructs hosting their light-responsive promoters. However, gene expression control with current or voltage would be more easily realized in hybrid whole-cell/microelectronic devices. A recent review summarizes the effects of electromagnetic fields (EMFs) and electric current pulses in living cells, including effects on gene expression [59]. Interestingly, while many of the most well-defined biological EMF effects have come from gene expression studies, much of this body of work has been difficult to reproduce. The most extensive series of investigations that implicate EMF effects on gene expression have been from the studies of Goodman and Henderson [45]. This group has published reports of *MYC*, β-actin, and histone H2B induction levels of up to threefold in HL60 cells after a 20-min exposure to several types of EMF signals [45]. However, rigorous attempts to repeat these experiments in independent laboratories have failed. Other researchers using nuclear run-on assays in T lymphoblastoid cells exposed to EMF for 15–120 min have reported induction of *Fos, Jun*, and *MYC* transcription [81]. However, in a similar study of *Fos, Jun*, and *JunB* genes in murine hematopoietic progenitor cell line, changes in gene expression were not detectable [84].

Additional reports relating the effect of electric current on gene expression have emerged from electroporation studies. This procedure, which entails exposing living cells to a single high-voltage electric pulse, induces the formation of transient membrane pores through which large biomolecules can pass. Originally developed for introducing exogenous DNA into cells, electroporation has also been reported to result in differential gene expression in a number of systems including interleukin-10 expression in monocytic cells [62]. Evidence for transcriptional modulation has also been reported in cells exposed to super-conducting magnets. In response to very high magnetic fields, an increase in expression of the *rpoS* gene in *E. coli* was observed [109].

Most recently electrical induction of *hsp70* gene expression has been demonstrated in murine astroglia and fibroblast cells. In this series of experiments murine cells were transfected with a reporter plasmid constructed by inserting the hsp70 promoter in front of a firefly luciferase gene. Subsequent electrical stimulation of the transfected cells resulted in light emission from the reporter [123]. Although very exciting, it is clear that the body of work in this area is sometimes inconsistent and even contradictory. These discrepancies have, in part, been attributed to lack of control of experimental variables relating to EMF parameters such as frequency, amplitude, and duration [10].

With the objective of identifying promoters for use in genetically engineered, electrically controllable whole-cell devices, we performed a search for current-inducible promoters in *Bacillus subtilis*. This particular organism was targeted both because its genome is fully sequenced and because, as spore former, it offers the potential for long-term storage when interfaced with electronic devices. In these experiments, all 4107 expressed genes in *B. subtilis* were screened for putative current-inducible genes. Cells were grown to log phase, placed in 35 mm × 7 mm electroporation cells (1.5 ml volume), and subjected to a current of 5 mA for a period of 10 min. An identical volume of cells was placed in buffer without electric current exposure. Simultaneously, cells were diluted and plated on nutrient agar plates to ascertain if the current had any lethal effects on cells. After current exposure, the induced and untreated cells were lysed and the RNA component isolated and quantified. Both RNA preparations were reverse transcribed incorporating ^{33}P-dATP to make the labeled cDNA that was hybridized against two identical commercial *B. subtilis* gene arrays (Genosys Panorama, The Woodlands, TX). These arrays permit the levels of all the expressed genes to be quantitatively assayed simultaneously.

None of the cells in the three treatment groups showed signs of lethality compared with controls. Based on statistical analysis of the 4 replicate arrays, 20 candidate genes were chosen as significantly differentially expressed. To verify differential expression, current-induced and control RNA was electrophoresed, blotted to nylon membranes, and hybridized with 300 basepair probes generated from each of the candidate genes by polymerase chain reaction (PCR). Based on Northern analysis, two genes were chosen for which *lux* reporter gene constructions would be made. The promoter sequences for these two genes were amplified from the genomic DNA by PCR. We are currently cloning these promoters into EZ:TN transposons that contain promoterless luciferase genes.

These reporter constructions will permit us to verify that the cloned promoters are indeed inducible by electric current or EMF. We cannot say at this point that these genes are specifically current inducible and not, for example, affected by osmotic or oxidative/reductive stress. From a pragmatic point of view, establishing specificity may not be necessary for the implementation of current-inducible promoters in engineered devices. From this perspective it is only necessary that the physiology of the cell is not unduly compromised during current induction and that the linked pathways function as designed.

ELECTROCHEMICAL CONTROL OF GENE REGULATION   Direct electrochemical generation or annihilation of chemical species may be implemented for chemical regulation of gene expression. Ultimately, such electrochemical interfacing could be implemented within individual cellular components, such that single cells might be addressed without diffusional cross-talk to neighboring cellular components. Approaches to such interfacing may follow the direction of electrochemical microelectrodes for cellular and subcellular chemical analyses that have been widely reported in the literature [117] and briefly described in a previous section of this chapter. In chapter 8 we address the construction of nanoscale electrochemical probes that could be used for this purpose.

DESIGN, SIMULATION, AND MODELING

The design of relatively simple systems can proceed in the absence of analysis, modeling, and simulation tools. Whole-cell biosensors relying on single reporter gene systems require only a phenomenological description of genetic circuit function. However, more complex devices that operate through the interaction of genetic circuits within an individual cell and, through cell-cell interconnectivity, between cells, cannot be designed easily without analytical models and simulation capabilities. The early efforts to engineer information transport within cells described above provide the groundwork for moving toward more complex engineered functionality of whole-cell components. Of particular interest, the toggle switch [42] and clock circuits [35] were designed with the aid of mathematical models, demonstrating progress toward analytical device models in the development and analysis of genetic circuits. While comparison of the predicted and observed circuit behavior indicates that these models are not complete, they represent a necessary step in the device science of cells, and continued efforts may provide insights into the complex functioning of natural genetic circuits. There are excellent reviews of cell modeling efforts [28, 36, 49, 70, 79, 83, 96].

At least for the near term, the modeling of genetic circuits is likely to differ somewhat from that found in the physical sciences, where models are based on fundamental concepts and contain a large number of parameters that can be individually measured [79]. However, the successful efforts to generate robust and predictive models of semiconductor devices and circuits can offer insights that may accelerate the development of analogous models for genetic circuits. It has been suggested that there is an element of convergent evolution between complex engineered devices and biological systems [23], and a rich history of electronic system design may provide clues for understanding gene network topology. A complete treatment of this material is beyond the scope of this chapter, but consider the following example.

Several strategies for analyzing or simulating the stochastic properties of genetic circuits have been reported [9, 47, 71, 78, 83, 102, 105]. Most often the results are given as signal-to-noise ratio, noise strength, stability parameters, or a time history of molecular concentration. Lost or hidden within such results are the frequency domain features of the noise, which are important as noise cascades through subsequent circuits. Maintaining the frequency domain features is especially important in autoregulated gene circuits analysis, as feedback impacts both magnitude and frequency composition of the noise [95]. A complete analysis requires that the frequency composition of the noise be preserved.

Recently we adapted frequency domain techniques borrowed from electronic circuit analysis that use the loop transmission concept to elucidate the noise performance of autoregulated (i.e., negative feedback) genetic circuits (Figure 5.8) while maintaining critical frequency information [95]. The loop transmission, T, is the transfer function around the loop and may be thought of as a measure of the resistance of the feedback loop to variation. T is calculated by introducing a perturbation (Δ) at any point within the circuit (e.g., a small change in transcription rate) and determining the response (ρ) that returns to the same point (e.g., a reactionary change in transcription rate). $T(f)$ is given by $\rho(f)/\Delta(f)$, and the feedback is negative if $T(0)$ is negative (i.e., has a phase of ± 180). For the autoregulated gene circuit in Figure 5.8b, $T(f)$ is given by [95]

$$T(f) = \frac{-T(0)}{\left(1 + i\dfrac{2\pi f}{\gamma_P}\right)\left(\dfrac{\pi f}{\gamma_R}\right)\left(1 + i\dfrac{f}{f_\alpha}\right)}$$

where b is the average number of proteins produced from each mRNA transcript, $\alpha(0)$ is the feedback term from a linear approximation to the Hill repression

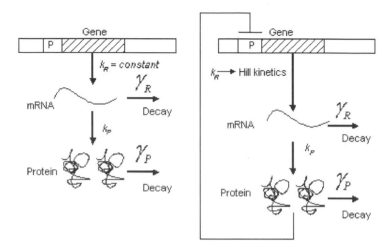

Figure 5.8 Model of single gene expression. (a) mRNA molecules are synthesized from the template DNA strand at rate k_R, and proteins are translated at a rate of k_P off of each mRNA molecule. γ_R and γ_P are the decay rates for mRNA and protein, respectively. (b) The same as panel a, except that the protein is negatively autoregulated. We model the autoregulation using a Hill expression such that $k_R = k_{R\,max}/(1 + (p/k_d)^n$, where $k_{R\,max}$ is the maximum rate of transcription, k_d is the concentration of protein where $k_R = k_{R\,max}/2$, and n is the Hill coefficient.

function, γ_p is the protein decay rate, γ_R is the mRNA decay rate, and f_α is a single-pole approximation of repression dynamics [95].

The frequency domain analysis of gene circuits like that shown in Figure 5.8b revealed two significant implications of negative autoregulation [95]. First, the noise strength (variance/mean) of the protein population in negatively autoregulated circuits is reduced by $1/[1 + |T(0)|]$ compared to open-loop gene circuits (Figure 5.8a). Second, autoregulation has an additional but more subtle effect on noise behavior that is not explicitly shown by other analyses. The noise bandwidth of the autoregulated circuit is increased by a factor of $1 + |T(0)|$ compared to the unregulated case. Thus, not only are the variance and standard deviation of the noise reduced by feedback, the noise that remains is shifted to higher frequencies (Figure 5.9).

It is possible that noise shifted to higher frequencies through the action of feedback may be filtered by downstream gene circuits [95]. There are electronic circuit-processing schemes that optimize performance by shifting noise into a frequency regime where it has a smaller impact on total system performance (e.g., sigma-delta analog-to-digital converters). Similar schemes could be used in engineered genetic circuits and may have evolved within natural genetic circuits.

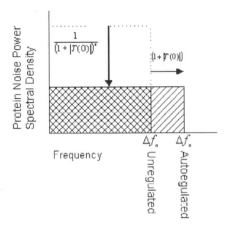

Figure 5.9 Graphical representation of the change in the protein noise power spectral density (PSD) caused by feedback. The magnitude of the PSD is reduced by a factor of $1/(1 + |T(0)|)^2$, but the bandwidth is increased by a factor of $1 + |T(0)|$. The net effect is a reduction of the noise strength by a factor of $1/(1 + |T(0)|)$.

Perhaps analysis tools like those presented above are best used to examine gene circuit structure–function relationships. These relationships may elucidate the selective pressures that influenced natural genetic circuit topology, or in synthetic biology applications, provide guidance in the selection of operational regimes to explore either through simulation or experiment. The frequency domain approach demonstrates the central role of loop transmission in noise behavior, the shifting of noise to higher frequency regimes through autoregulation, the trade-off between noise reduction through negative autoregulation and excessive high frequency noise as phase margin is diminished, and the noise optimizing ordering of parameters that control dynamics. This understanding of simpler components allows an examination of the structure–function relationship at higher levels of gene circuit ordering. Autoregulated gene circuits, like those used in the example above, almost certainly play an important role in gene networks [106], and the history of the development of feedback configurations for electronic systems may be a resource for developing an understanding of gene circuits and the selective pressure that drove their evolution. The extent to which this analogy is useful can only be determined by a great deal more analysis, modeling, and experimentation. Conversely, it is possible that new strategies for engineered system design may emerge from this examination. Certainly the development of electronic feedback amplifiers in the 1920s may have benefited from an understanding of autoregulated gene function, and it is possible that currently unidentified gene network topologies may impact future system design strategies.

CONCLUSIONS

Heretofore unknown mechanisms that living cells use to process multiple sensory signals from the environment, as well as other cells, are yielding to molecular investigation. As knowledge on these mechanisms is accumulated, it is possible to synthesize via genetic engineering chimeras capable of directed information processing with electronic hybrid functionality. Such chimeras are natural complements to man-made information processing systems and are the basis for a new generation of microscale and nanoscale sensors, machines, and computational systems.

The simple one gene or one enzyme yielding one response paradigm that cells use in information processing has given way to a vision of biocomplexity that was unappreciated until recently. However, it is clear that this complexity is not unmanageable. The complexity demonstrated by higher ordered eukaryotic cells can be readily mimicked in lower eukaryotic and prokaryotic cells, which may provide robust components in nano- and microscale devices of the future. Systematic attempts to model cellular functions at the level of metabolism, gene expression, and signal transduction have provided insight into fruitful paths of experimentation to create microorganisms that carry out rudimentary information processing among cells, with bidirectional communication with a synthetic information-processing system. It appears that direct communication with the cells to initiate signal recognition events and information processing is at hand. The combination of these two states realizes hybrid devices capable of environmental awareness with decision-making ability to respond to or control that immediate environment.

The eventual application for these devices must arise from the natural advantages of information processing within whole cells, while minimizing the disadvantages associated with living device components. High-speed computation will not be the strong suit of these devices. However, analysis, actuation, and control of highly complex situations and environments with time constants similar to those of the cell's reaction-diffusion signal transport may be well served by these hybrid devices. Applications may include therapeutic drug discovery; biomedical devices for disease diagnosis, treatment, or long-term management; industrial chemical process control; environmental monitoring and control; and management of closed environments, such as long-term, manned space flight compartments.

An application not to be overlooked is in the fundamental science of unravelling natural biological complexity. There is a growing shift in emphasis away from purely reductionistic approaches toward more integrative "cells as systems" approaches [79]. At least initially, this fundamental science and the device science will travel parallel and synergistic paths. Looking to the future, design tools in the form analysis and modeling approaches that reliably predict the behavior of whole-cell components are required to enable the rational engineering of these hybrid systems. Such tools are considered indispensable in the analogous field of electronic circuit design, and they are likely to play a similar role in the design of hybrid systems that include whole cell components.

The frequency domain gene circuit analysis techniques described here illustrate an important connection between insights developed through more than a century of electronic circuit design and the currently developing understanding of the architecture of naturally occurring genetic circuits. These insights will aid both in the design of synthetic gene circuits and in the elucidation of naturally occurring gene circuits, leading to an infrequently encountered parallel and simultaneous development of both an engineering and a basic science discipline. It is even possible that these efforts will go full circle with new electronic circuit topologies arising from a deeper understanding of the structure of gene circuits.

Although recent progress has been encouraging, and in some cases even remarkable, this field is very much in its infancy. The scientific and technological challenges are significant, but educational and research cultural issues loom just as large. The full breadth of this endeavor encompasses cell, molecular, and microbiology; electrical, computer, and material science and engineering; chemistry, physics, and mathematics. Graduate education and research programs will need to become more integrated and flexible to promote the training of new research professionals to address these issues. Furthermore, peer review panels will have to become more tolerant of multi- and interdisciplinary research proposals to allow funding to flow into these areas. With these changes, an entirely new discipline will eventually emerge, and a new technology will evolve into a practical reality.

References

[1] B. B. Anderson and A. G. Ewing. Chemical profiles and monitoring dynamics at an individual nerve cell in planorbis corneus with electrochemical detection. *J. Pharm. Biomed. Anal.*, 19:15–32, 1999.

[2] J. Ross Arkin and H. H. McAdams. Stochastic kinetic analysis of developmental pathway bifurcation in phage lambda infected *Escherichia coli* cells. *Genetics*, 149:1633–1648, 1998.

[3] D. L. Balkwill, D. Maratea, and R. P. Blakemore. Ultrastructure of a magnetotactic spirillum. *J. Bacteriol.*, 141:1399–1408, 1980.

[4] B. L. Bassler. A multichannel two-component signaling relay controls quorum sensing in *Vibrio harveyi*. In G.M. Dunny and S.C. Winans, editors, *Cell-Cell Signaling in Bacteria*, pages 259–273. ASM Press, Washington, DC, 1999.

[5] B. L. Bassler, E. P. Greenberg, and A. M. Stevens. Cross-species induction of luminescence in the quorum-sensing bacterium *Vibrio harveyi*. *J. Bacteriol.*, 179:4043–4045, 1997.

[6] D. A. Bazylinski et al. Controlled biomineralization of magnetite (fe3o4) and greigite (fe3s4) in a magnetotactic bacterium. *Appl. Environ. Microbiol.*, 61: 3232–3239, 1995.

[7] D. A. Bazylinski, R. B. Frankel, and H. W. Jannasch. Anaerobic magnetite production by a marine, magnetotactic bacterium. *Nature*, 334:518–519, 1988.

[8] A. Becskei and L. Serrano. Engineering stability in gene networks by autoregulation. *Nature*, 405:590–593, 2000.

[9] A. Becskei and L. Serrano. Engineering stability in gene networks by autoregulation. *Nature*, 405:590–593, 2000.

[10] H. Berg. Problems of weak electromagnetic field effects in cell biology. *Bioelectrochem. Bioenergetics*, 48:355–360, 1999.

[11] P. Billard and M. S. DuBow. Bioluminescence-based assays for detection and characterization of bacteria and chemicals in clinical laboratories. *Clin. Biochem.*, 31:1–14, 1998.

[12] R. P. Blakemore. Magnetotactic bacteria. *Science*, 190:377–379, 1975.

[13] F. Blanche et al. Purification and characterization of S-adenosyl-L-methionine: uroporphyrinogen iii methyltransferase from *Pseudomonas trificans*. *J. Bacteriol.*, 171:4222–4231, 1989.

[14] F. Blanche et al. Purification, characterization, and molecular cloning of S-adenosyl-L-methionine: uroporphyrinogen iii methyltransferase from *Methanobacterium ivanovii*. *J. Bacteriol.*, 173:4637–4645., 1991.

[15] N. Bollinger et al. Gene expression in *Pseudomonas aeruginosa*: evidence of iron override effects on quorum sensing and biofilm-specific gene regulation. *J. Bacteriol.*, 183:1990–1996, 2001.

[16] M. R. W. Brown and P. Gilbert. Sensitivity of biofilms to antimicrobial agents. *J. Appl. Bacteriol.*, 74:S87–S97, 1993.

[17] M. J. Casadaban and S. N. Cohen. Lactose genes fused to exogenous promoters in one step using a mu-lac bacteriophage: in vivo probe for transcriptional control sequences. *Proc. Natl. Acad. Sci. USA*, 76:4530–4533, 1979.

[18] D. J. Christini, J. Walden, and J. M. Edelberg. Direct biologically based biosensing of dynamic physiological function. *Am. J. Physiol.-Heart Circul. Physiol.*, 280(5):H2006–10, 2001.

[19] R. A. Clark, S. E. Zerby, and A. G. Ewing. Electrochemistry in neuronal microenvironments. *Electroanal. Chem.*, 20:227–295, 1998.

[20] J. W. Costerton et al. Bacterial biofilms in nature and disease. *Annu. Rev. Microbiol.*, 41:435–464, 1987.

[21] J. W. Costerton et al. Microbial biofilms. *Annu. Rev. Microbiol.*, 49:711–745, 1995.

[22] J. W. Costerton, P. S. Stewart, and E. P. Greenberg. Bacterial biofilms: a common cause of persistent infections. *Science*, 284:1318–1322, 1999.

[23] M. E. Csete and J. C. Doyle. Reverse engineering of biological complexity. *Science*, 295:1664–1669, 2002.

[24] P. N. Danese, L. A. Pratt, and R. Kolter. Biofilm formation as a developmental process. *Methods Enzymol.*, 336A:19–28, 2001.

[25] S. Daunert et al. Genetically engineered whole-cell sensing systems: coupling biological recognition with reporter genes. *Chem. Rev.*, 100(7):2705–2738, 2000.

[26] M. E. Davey and G. O'Toole. Microbial biofilms: from ecology to molecular genetics. *Microbiol. Mol. Biol. Rev.*, 64:847–867, 2000.

[27] D. G. Davies et al. The involvement of cell-to-cell signals in the development of a bacterial biofilm. *Science*, 280:295–298, 1998.

[28] H. De Jong. Modeling and simulation of genetic regulatory networks: a literature review. *J. Comput. Biol.*, 9:67–103, 2002.

[29] T. R. de Kievit et al. Quorum-sensing genes in pseudomonas aeruginosa biofilms: their role and expression patterns. *Appl. Environ. Microbiol.*, 67:1865 1873, 2001.

[30] T. R. de Kievit and B. H. Iglewski. Bacterial quorum sensing in pathogenic relationships. *Infect. Immun.*, 68:4839–4849, 2000.

[31] Alvarez de Toledo, R. Fernandez-Chacon G., and J. M. Fernandez. Release of secretory products during transient vesicle fusion. *Nature*, 363.554–558, 1993.

[32] B. D. DeBusschere and G. T. A. Kovacs. Portable cell-based biosensor system using integrated cmos cell-cartridges. *Biosens. Bioelectr.*, 16:543–556, 2001.

[33] J. DeRisi et al. Use of a cDNA microarray to analyse gene expression patterns in human cancer. *Nat. Genet.*, 14:457–460, 1996.

[34] J. DeRisi et al. Genome microarray analysis of transcriptional activation in multidrug resistance yeast mutants. *FEBS. Lett.*, 470:156–160, 2000.

[35] M. B. Elowitz and S. Leibler. A synthetic oscillatory network of transcriptional regulators. *Nature*, 403:335–338, 2000.

[36] D. Endy and R. Brent. Modelling cell behaviour. *Nature*, 409:391–395, 2001.

[37] M. Farina, D. M. S. Esquivel, and H. G. P. Lins de Barros. Magnetic iron-sulfur crystals from a magnetotactic microorganism. *Nature*, 334:256–258, 1990.

[38] R. B. Frankel. Magnetic guidance of organisms. *Annu. Rev. Biophys. Bioeng.*, 13:85–103, 1984.

[39] R. B. Frankel, R. P. Blakemore, and R. S. Wolfe. Magnetite in freshwater magnetotactic bacteria. *Science*, 203:1355–1356, 1979.

[40] P. Fromherz and A. Stett. Silicon-neuron junction: capacitive stimulation of an individual neuron on a silicon chip. *Phys. Rev. Lett.*, 75:1670–1673, 1995.

[41] C. Fuqua and E. P. Greenburg. Self perception in bacteria: quorum sensing with acylated homoserine lactones. *Curr. Opin. Microbiol.*, 1(2):183–189, 1998.

[42] T. S. Gardner, C. R. Cantor, and J. J. Collins. Construction of a genetic toggle switch in *Escherichia coli*. *Nature*, 403:339–342, 2000.

[43] D. T. Gillespie. Exact stochastic simulation of coupled chemical-reactions. *J. Phys. Chem.*, 81:2340–2361, 1977.

[44] B. S. Goldman and J. R. Roth. Genetic structure and regulation of the cysg gene in salmonella typhimurium. *J. Bacteriol.*, 175:1457–1466, 1993.

[45] R. Goodman, L. X. Wei, J. Bumann, and A. Shirely-Henderson. Exposure to electric and magnetic fields increase transcripts in hl-60 cells: does adaptation to em fields occur. *Bioelectrochem. Bioenerg.*, 29:185–192., 1992.

[46] K. M. Gray, L. Passador, B. H. Iglewski, and E. P. Greenberg. Interchangeability and specificity of components from the quorum-sensing regulatory systems of vibrio fischeri and pseudomonas aeruginosa. *J. Bacteriol.*, 176:3076–3080, 1994.

[47] E. P. Greenberg. Acyl-homoserine lactone quorum sensing in bacteria. *J. Microbiol.*, 38:117–121, 2000.

[48] G. W. Gross and F. U. Schwalm. A closed chamber for long-term electrophysiological and microscopical monitoring of monolayer neuronal networks. *J. Neurosci. Meth.*, 52:73–85, 1994.

[49] J. Hasty, D. McMillen, F. Isaacs, and J. J. Collins. Computational studies of gene regulatory networks: In numero molecular biology. *Nature Rev. Genet.*, 2(4):268–279, 2001.

[50] B. R. Heywood et al. Controlled biosynthesis of greigite (Fe_3S_4) in magnetotactic bacteria. *Naturwissenschaften*, 77(11):536–538, 1990.

[51] M. Hiratsuka, T. Aoki, and T. Higuchi. Enzyme transistor circuits for reaction-diffusion computing. *IEEE Trans. Cir. Sys. I Fundam. Theory Appl.*, 46(2):294–303, 1999.

[52] M. Hiratsuka, T. Aoki, and T. Higuchi. Pattern formation in reaction-diffusion enzyme transistor circuits. *IEICE Trans. Fundam. Electron. Commun. Computer Sci.*, E82(9):1809–1817, 1999.

[53] L. Huang and R. T. Kennedy. Exploring single-cell dynamics using chemically modified electrodes. *Trends Anal. Chem.*, 14(4):158–164, 1995.

[54] ITRS. International Technology Roadmap for Semiconductors, available at http://public.itrs.net/.

[55] V. R. Iyer et al. The transcriptional program in the response of human fibroblasts to serum. *Science*, 283:83–87, 1999.

[56] M. Jia et al. Global expression profiling of yeast treated with an inhibitor of amino acid biosynthesis, sulfometuron methyl. *Physiol. Genom.*, 3:83–92, 2000.

[57] J. M. Kendall and M. N. Badminton. *Aequorea victoria* bioluminescence moves into an exciting new era. *Trends Biotechnol.*, 16:216–224, 1998.

[58] C. J. Kenyon and G. C. Walker. DNA-damaging agents stimulate gene expression at specific loci in *Escherichia coli*. *Proc. Natl. Acad. Sci. USA*, 77:2819–2823, 1980.

[59] A. Lacy-Hulbert, J. C. Metcalfe, and R. Hesketh. Biological response to electromagnetic fields. *FASEB J.*, 12:395–420, 1998.

[60] R. A. LaRossa and T. K. Van Dyk. Applications of stress responses for environmental monitoring and molecular toxicology. In G. Storz and R. Hengge-Aronis, editors, *Bacterial Stress Responses*, pages 455–468. ASM Press, Washington, DC, 2000.

[61] Y. Y. Lau, T. Abe, and A. G. Ewing. Voltammetric measurement of oxygen in single neurons using platinized carbon ring electrodes. *Anal. Chem.*, 64:1702–1705., 1992.

[62] M. H. Lehmann and H. Berg. Electroporation induced gene expression-a case study on interleukin-10. *Bioelctrochem. Bioenerg.*, 47:3–10, 1998.

[63] H. S. Liu et al. Is green fluorescent protein toxic to the living cells? *Biochem. Biophys. Res. Commun.*, 260:712–717, 1999.

[64] H. Lu and M. Gratzl. Monitoring drug efflux from sensitive and multidrug-resistant single cancer cells with microvoltammetry. *Anal. Chem.*, 71:2821–2830, 1999.

[65] S. Mann et al. Biomineralization of ferrimagnetic greigite (Fe_3S_4) and iron pyrite (FeS_2) in a magnetotactic bacterium. *Nature*, 343:258–261, 1990.

[66] S. Mann, N. H. C. Sparks, and R. P. Blakemore. Ultrastructure and characterization of anisotropic magnetic inclusions in magnetotactic bacteria. *Proc. R. Soc. Lond. B*, 231:469–477, 1987.

[67] M. Marton et al. Drug target validation and identification of secondary drug target effects using dna microarrays. *Nat. Med.*, 4:1293–1301, 1998.

[68] T. Matsuda et al. Morphology and structure of biogenic magnetite particles. *Nature*, 302:411–412, 1983.

[69] H. H. McAdams and A. Arkin. Stochastic mechanisms in gene expression. *Proc. Natl. Acad. Sci. USA*, 94:814–819, 1997.

[70] H. H. McAdams and A. Arkin. Simulation of prokaryotic genetic circuits. *Annu. Rev. Biophys. Biomol. Struct.*, 27:199–224., 1998.

[71] H. H. McAdams and A. Arkin. Stochastic mechanisms in gene expression. *Proc. Natl. Acad. Sci. USA*, 94:814–819, 1997.

[72] E. Meighen. Bioluminescence, bacterial. *Encycl. Microbiol.*, 1:309–319, 1992.

[73] E. A. Meighen. Molecular biology of bacterial bioluminescence. *Microbiol. Rev.*, 55:123–142, 1991.

[74] L. H. Naylor. Reporter gene technology: the future looks bright. *Biochem. Pharmacol.*, 58:749–757, 1999.

[75] J. C. Nickel et al. Tobramycin resistance of *Pseudomonas aeruginosa* cells growing as a biofilm on urinary catheter material. *Antimicrob. Agents Chemother.*, 27(4):619–624, 1985.

[76] S. V. Nyholm et al. Establishment of an animal-bacterial association: recruiting symbiotic vibrios from the environment. *Proc. Natl. Acad. Sci. USA*, 97(18): 10231–10235, 2000.

[77] G. O'Toole, H. B. Kaplan, and R. Kolter. Biofilm formation as microbial development. *Annu. Rev. Microbiol.*, 54:49–79., 2000.

[78] E. M. Ozbudak, M. Thattai, I. Kurtser, A. D. Grossman, and A. can Oudenaarden. Regulation of noise in the expression of a single gene. *Nature Genetics*, 31:69–73, 2002.

[79] B. Palsson. The challenges of in silico biology. *Nature Biotechnol.*, 18:1147–1150, 2000.

[80] J. J. Pancrazio et al. Development and application of cell-based biosensors. *Ann. Biomed. Eng.*, 27:697–711.

[81] J. L. Philips et al. Magnetic field induced changes in specific gene transcription. *Biochim. Biophys. Acta*, 1132:140–144, 1992.

[82] G. N. Phillips Jr. Structure and dynamics of green fluorescent protein. *Curr. Opin. Struct. Biol.*, 7:821–827, 1997.

[83] C. Rao, D. M. Wolf, and A. P. Arkin. Control, exploitation and tolerance of intracellular noise. *Nature*, 420:231–237, 2002.

[84] B. M. Reipert, D. Allan, and T. M. Dexter. Exposure to extremely low frequency

magnetic fields has no effect on growth rate or clonogenic potential of multipotential haemopoietic progenitor cells. *Growth Factors*, 13:205–217, 1996.

[85] C. S. Richmond et al. Genome-wide expression profiling in *Escherichia coli* k-12. *Nucleic Acids Res.*, 27:3821–3835, 1999.

[86] C. Robin et al. Primary structure, expression in *Escherichia coli*, and properties of S-adenosyl-L-methionine:uroporphyrinogen iii methyltransferase from *Bacillus megaterium. J. Bacteriol.*, 173:4893–4896, 1991.

[87] C. A. Roessner and A. I. Scott. Fluorescence-based method for selection of recombinant plasmids. *Biotechniques*, 19:760–764, 1995.

[88] E. G. Ruby. Lessons from a cooperative, bacterial-animal association: the *Vibrio fischeri Euprymna scolopes* light organ symbiosis. *Annu. Rev. Microbiol.*, 50:591–624, 1996.

[89] K. P. Rumbaugh, J. A. Griswold, and A. N. Hamood. The role of quorum sensing in the in vivo virulence of *Pseudomonas aeruginosa. Microb. Infect.*, 2:1721–1731, 2000.

[90] I. Sattler et al. Cloning, sequencing, and expression of the uroporphyrinogen iii methyltransferase coba gene of *Propionibacterium freudenreichii (shermanii). J. Bacteriol.*, 177:1564–1569, 1995.

[91] J. T. Shingleton et al. Induction of the tod operon by trichloroethylene in *Pseudomonas putida* tva8. *Appl. Environ. Microbiol.*, 64(12):5049–5052, 1998.

[92] T. J. Silhavy. Gene fusions. *J. Bacteriol.*, 182:5935–5938, 2000.

[93] M. L. Simpson et al. Application specific spectral response with cmos compatible photodiodes. *IEEE Trans. Elect. Dev.*, 46(5):905–913, 1999.

[94] M. L. Simpson, G. S. Sayler, J. T. Fleming, and B. A. Applegate. Whole-cell biocomputing: engineering the information processing functionality of cells. *Trends Biotechnol.*, 19(8):317–323, 2001.

[95] M. L. Simpson, C. D. Cox, and G. S. Sayler. Frequency domain analysis of noise in autoregulated gene circuits. *Proc. Natl. Acad. Sci. USA*, 100:4551–4556, 2003.

[96] P. Smolen, D. A. Baxter, and J. H. Byrne. Modeling transcriptional control in gene networks - methods, recent results, and future directions. *Bull. Math. Biol.*, 62:247–292, 2000.

[97] D. A. Stenger et al. Detection of physiologically active compounds using cell-based biosensors. *Trends Biotech.*, 19(8):304–309, 2001.

[98] A. Stett, B. Muller, and P. Fromherz. Two-way silicon-neuron interface by electrical induction. *Phys. Rev. E*, 55(2):1779–1782, 1997.

[99] A. M. Stevens and E. P. Greenberg. Quorum sensing in *Vibrio fischeri*: essential elements for activation of the luminescence genes. *J. Bacteriol.*, 179:557–562, 1997.

[100] A. Stintzi, K. Evans, J. M. Meyer, and K. Poole. Quorum-sensing and siderophore biosynthesis in *Pseudomonas aeruginosa*: lasr/lasi mutants exhibit reduced pyoverdine biosynthesis. *FEMS Microbiol. Lett.*, 166:341–345, 1998.

[101] M. G. Surette and B.L. Bassler. Quorum sensing in *Escherichia coli* and *Salmonella typhimurium. Proc. Natl. Acad. Sci. USA*, 95:7046–7050, 1998.

[102] P. S. Swain, M. B. Elowitz, and E. D. Siggia. Intrinsic and extrinsic contributions to stochasticity in gene expression. *Proc. Natl. Acad. Sci. USA*, 99:12795–12800, 2002.

[103] S. Swift, P. Williams, and G.S. Stewart. N-acylhomoserine lactones quorum sensing in proteobacteria. In G.M. Dunny and S.C. Winans, editors, *Cell-Cell Signaling in Bacteria*, pages 291–313. ASM Press, Washington, DC, 1999.

[104] H. Tao et al. Functional genomics: expression analysis of *Escherichia coli* growing on minimal and rich media. *J. Bacteriol.*, 181:6425–6440, 1999.

[105] M. Thattai and A. van Oudenaarden. Intrinsic noise in gene regulatory networks.
Proc. Natl. Acad. Sci. USA, 98:8614–8619, 2001.

[106] D. Thieffry, A.M. Huerta, E. Perez-Rueda, and J. Collado-Vides. From specific
gene regulation to genomic networks: a global analysis of transcriptional regulation in escherichia coli. *Bioessays*, 20:433–440, 1998.

[107] K. M. Towe and T. T. Moench. Electron-optical characterization of bacterial
magnetite. *Earth Planet. Sci. Lett.*, 52:213–220, 1981.

[108] R. Y. Tsien. The green fluorescent protein. *Annu. Rev. Biochem.*, 67:509–544,
1998.

[109] K. Tsuchiya et al. High magnetic field enhances stationary phase-specific transcription activity of *Escherichia coli*. *Bioelectrochem. Bioenergetics*, 48:383–
387, 1999.

[110] T. K. Van Dyk et al. A genomic approach to gene fusion technology. *Proc. Natl.
Acad. Sci. USA*, 98(5):2555–2560, 2001.

[111] K. L. Visick et al. *Vibrio fischeri* lux genes play an important role in colonization
and development of the host light organ. *J. Bacteriol.*, 182(16):4578–4586, 2000.

[112] B. L. Wanner and R. McSharry. Phosphate-controlled gene expression in *Escherichia coli* k12 using mudl-directed lacz fusions. *J. Mol. Biol.*, 158:347–363,
1982.

[113] M. J. Warren, C. A. Roessner, P. J. Santander, and A. I. Scott. The *Escherichia
coli* cysg gene encodes S-adenosylmethionine-dependent uroporphyrinogen iii
methylase. *Biochem. J.*, 265:725–729, 1990.

[114] R. Weiss, G. Homsy, and T. F. Knight. Toward in vivo digital circuits. In Laura
Landweber and Erik Winfree, editors, *Evolution as Computation*, pages 275–295.
Springer, Berlin, 2003..

[115] R. Weiss and T. F. Knight Jr. Engineered communications for microbial robotics.
Proc. 6th International Workshop on DNA-Based Computers, Lecture Notes in
Computer Science, vol. 2054, pages 1–16. Springer-Verlag, Berlin, Heidelberg,
2001.

[116] S. Welsh and S. A. Kay. Reporter gene expression for monitoring gene transfer.
Curr. Opin. Biotechnol., 8:617–622, 1997.

[117] R. M. Wightman et al. Temporally resolved catecholamine spikes correspond to
single vesicle release from individual chromaffin cells. *Proc. Natl. Acad. Sci.
USA*, 88:10754–10758, 1991.

[118] S. Wildt and U. Deuschle. Coba, a red fluorescent transcriptional reporter for
Escherichia coli, yeast, and mammalian cells. *Nat. Biotechnol.*, 17:1175–1178,
1999.

[119] P. Williams et al. Quorum sensing and the population-dependent control of virulence. *Phil. Trans. R. Soc. Lond. B*, 355:667–680, 2000.

[120] T. Wilson and J. W. Hastings. Bioluminescence. *Annu. Rev. Cell. Dev. Biol.*,
14:197–230, 1998.

[121] H. Withers, S. Swift, and P. Williams. Quorum sensing as an integral component of gene regulatory networks in gram-negative bacteria. *Curr. Opin. Microbiol.*, 4:186–193, 2001.

[122] K. V. Wood. Marker proteins for gene expression. *Curr. Opin. Biotechnol.*, 6:50–58, 1995.

[123] Y. Yanagida et al. Electrically stimulated induction of hsp70 gene expression in mouse astroglia and fibroblast cells. *J. Biotechnol.*, 79:53–61, 2000.

[124] G. Zeck and P. Fromherz. Noninvasive neuroelectronic interfacing with synaptically connected snail neurons immobilized on a semiconductor chip. *Proc. Natl. Acad. Sci. USA*, 98(18):10457–10462, 2001.

Part II

Laboratory Experiments

6

The *Enterococcus faecalis* Information Gate

Kenichi Wakabayashi and
Masayuki Yamamura

Information exchange between cellular compartments allows us to engineer systems based around cooperative principles. In this chapter we consider a unique bacterial communication system, the conjugative plasmid transfer of *Enterococcus faecalis*. Using these bacteria, we describe how to engineer a logically controlled information gate and build a logical inverter based upon it.

INTRODUCTION

Cellular computing is an alternative computing paradigm based on living cells [1]. Microscale organisms, especially bacteria, are well suited for computing for several reasons. A small culture provides an almost limitless supply of bacterial "hardware." Bacteria can be stored and easily modified by gene recombination. In addition, and important for our purposes, bacteria can produce various signal molecules that are useful for computation.

DNA-binding proteins recognize specific regulatory regions of DNA, bind them, and regulate their genetic expression. These proteins are available for use as computing signals inside the cell. Weiss et al. [8] have shown, for example, how to construct logic circuits based on gene expression regulated by DNA-binding proteins (see also chapter 4).

Some signal molecules are associated with intercellular communications between individuals. Intercellular communication is one of the fundamental characteristics of multicellular organisms, but it is also found in single-celled microorganisms, including bacteria. Communication mediated by homoserine lactones can widely be seen in various Gram-negative bacteria. The mechanism of this behavior was well characterized in *Vibrio fischeri*, due to their bioluminescent activity mediated by homoserine lactones [6]. It has been shown that bacterial information transfer can be engineered as an extension of *Escherichia coli* into which the *lux* genes of *Vibrio fischeri* are transformed [9]. The communication abilities of bacteria therefore allow us to build microbial information processors for cellular computing.

PHEROMONE-INDUCED CONJUGATION AND PLASMID TRANSFER OF *ENTEROCOCCUS FAECALIS*

Communication mechanisms in Gram-positive bacteria are not yet well understood. One of the exceptions to this is the conjugative plasmid transfer system in *Enterococcus faecalis* [3, 4, 10]. *E. faecalis* conjugate in response to a pheromone is released by other cells (Figure 6.1). Pheromones are seven- or eight-residue amino peptides produced in *E. faecalis* (Table 6.1). In the case of cPD1, the pheromone is produced by truncation of a 22-residue precursor that is the signal peptide of a lipoprotein. Pheromones are secreted to the outside of the cell and received by other *E. faecalis* cells that carry the conjugative plasmid. The pheromones activate transcription of plasmid genes that encode particular cell-surface adhesion molecules. Expression of the adhesion molecules on the cell surface induces cell aggregation in *E. faecalis*. Once cell aggregation occurs, the plasmid is transferred from cell to cell, one cell acting as the donor and another as the recipient during conjugative plasmid transfer.

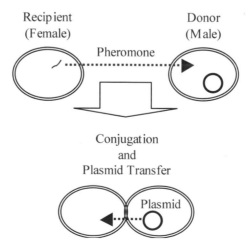

Figure 6.1 Pheromone-induced conjugative plasmid transfer in *E. faecalis*.

Table 6.1 Structures of pheromones and inhibitors.

Pheromon/inhibitor	Peptide structure
cPD1	H-Phe-Leu-Val-Met-Phe-Leu-Ser-Gly-OH
iPD1	H-Ala-Leu-Ile-Leu-Thr-Leu-Val-Ser-OH
cAD1	H-Leu-Phe-Ser-Leu-Val-Leu-Ala-Gly-OH
iAD1	H-Leu-Phe-Val-Val-Thr-Leu-Val-Gly-OH
cCF10	H-Leu-Val-Thr-Leu-Val-Phe-Val-OH
iCF10	H-Ala-Ile-Thr-Leu-Ile-Phe-Ile-OH
cAM373	H-Ala-Ile-Phe-Ile-Leu-Ala-Ser-OH
iAM373	H-Ser-Ile-Phe-Thr-Leu-Val-Ala-OH
cOB1	H-Val-Ala-Val-Leu-Val-Leu-Gly-Ala-OH
iOB1	H-Ser-Leu-Thr-Leu-Ile-Leu-Ser-Ala-OH

Conjugative Plasmids, Pheromones, and Inhibitors

Many conjugative plasmids are found in various *E. faecalis* strains (Table 6.2). These plasmids are classified by phenotype in terms of the type of pheromone to which they can respond. A plasmid is responsive only to its corresponding class of pheromone.

In the free-living state, *E. faecalis* constantly produce pheromones of every class, unless they carry a conjugative plasmid. When they do carry a conjugative plasmid, corresponding pheromone production is reduced by the specific regulatory mechanism directed by the genes on the conjugative plasmid (*pheromone shutdown*).

In addition, *E. faecalis* that carry a conjugative plasmid produce a pheromone inhibitor. The inhibitor gene is located on the plasmid. Pheromone inhibitors are all seven- or eight-residue amino peptides. The amino acid sequences of inhibitors are partially homologous to that of pheromones of the same class (Table 6.1). Pheromone inhibitors competitively bind the pheromone receptor and inhibit the downstream pathway of cell activation induced by the pheromone.

IMPLEMENTATION OF THE INFORMATION GATE

The conjugation system of *E. faecalis* is applicable to building information processors for cellular computing. Pheromones serve as signals, and the information written in plasmids is transferred from one cell to another.

E. faecalis Information Gate

Using combined inhibitors, we designed a logically controlled DNA transporter (Figure 6.2). Pheromone C_i and inhibitor I_i are used as gate signals. Depending

Table 6.2 Conjugative plasmids of *E. faecalis*.

Plasmid	Size (kbp)	Phenotype[a]
pAD1	60	cAD1 responsive, Hly/Bac
pAM323	66	cAM323 responsive, Em^r
pAM324	53	cAM324 responsive
pAM373	36	cAM373 responsive
pAMγ2	~60	cAMγ2 responsive, Bac
pAMγ3	~60	cAMγ3 responsive
pCF10	54	cCF10 responsive, Tc^r
pIP1017	?	cIP1017 responsive, Km^r, Sm^r
pIP1141	?	cIP1141 responsive
pIP1438	?	cIP1438 responsive, Cm^r, Em^r
pIP1440	?	cIP1440 responsive, Tc^r, Sm^r
pJH2	59	cAD1 responsive, Hly/Bac
pMB1	?	cCF10 responsive
pMB2	?	cPD1 responsive
pMV120	?	cMV120 responsive
pOB1	71	cOB1 responsive, Hly/Bac
pPD1	56	cPD1 responsive, Bac
pYI2	?	cYI2 responsive, Hly/Bac
pYI7	?	cYI7 responsive, Bac

[a] Hly; hemolysin, Bac; bacteriocin, Tc^r; tetracycline resistance, Em^r; erythromycin resistance, Km^r; kanamycin resistance, Sm^r; streptmycin resistance, Cm^r; chloramphenicol resistance.

on the state of the input signals, the plasmid P_i is either released or not released from the gate. In the presence of C_i (and the absence of I_i), the gate releases P_i. The gate can be multiplied in parallel with a variety of the conjugative plasmids (Figure 6.3).

Conjugative Plasmid Transfer

The *E. faecalis* strains and plasmids used in this study are listed in Table 6.3. To unify the strains of *E. faecalis*, we transferred the conjugative plasmid pCF10 from *E. faecalis* strain OG1SSp to strain OG1RF and transferred plasmid pAM714 from strain OG1X to strain OG1RF. The procedure for plasmid

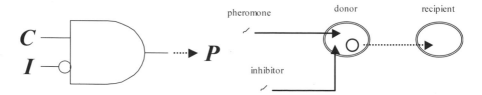

Figure 6.2 A primitive gate-unit of *E. faecalis* information gate.

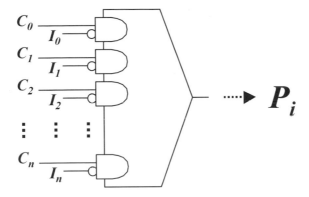

Figure 6.3 Multiplied *E. faecalis* information gate.

transfer is described in Dunny et al. [5]. OG1SSp(pCF10) and OG1RF were cultured in 1 ml of THB (Todd-Hewitt broth) at 37°C for 18 h. We added 10 μg tetracycline to the culture medium of OG1SSp(pCF10) to retain the plasmid in the host cell. After incubation, the culture soup of OG1SSp and OG1RF was removed completely. We combined OG1SSp(pCF10) and OG1RF after washing twice with fresh THB. The cell mixtures were resuspended in the previously removed OG1RF soup. The cell mixtures were incubated at 37°C for 2.5 h, and, after incubation, were spread on a THB agar plate containing 10 μg/ml tetracycline and 50 μg/ml fusidic acid. The cells on the plates were incubated at 37°C for 18 h. We picked a bacterial colony of OG1RF(pCF10) and cultured it in 1 ml THB with tetracycline and fusidic acid. We added 1 ml glycerol to the bacterial culture, which was then stocked at 80°C. We obtained OG1RF(pAM714) in a similar manner.

Table 6.3 Strains and plasmids used in this work.

E. faecalis *strain*	*Phenotype*[a]
OG1X	Sm^r
OG1RF	Rf^r, Fa^r
OG1SSp	Sm^r, Sp^r
Conjugative plasmid	
pCF10	Tc^r, cCF10 responsive
pAM714	Em^r, cAD1 responsive
pAM351	Tc^r, cPD1 responsive
pOB1 (::Tc^r)	Tc^r, cOB1 responsive

[a]Sm^r; streptmycin resistance, Rf^r; refampicin resistance, Fa^r; fusidic acid resistance, Sp^r; spectinomycin resistance, Tc^r; tetracycline resistance, Em^r; erythromycin resistance.

Realization of the E. faecalis *Information Gate*

E. faecalis secrete various pheromones into the culture medium. Even an hour-long culture of *E. faecalis* is sufficient to induce *E. faecalis* conjugative plasmid transfer. *E. faecalis* donors also secrete pheromone inhibitors. However, the amount of secretion is not sufficient for our purposes, so we used synthetic pheromone peptides in this work.

We incubated OG1RF(pAM714) and OG1SSp in THB at 37°C for 18 h. After washing twice with fresh THB, OG1RF(pAM714) was incubated for 60 min in THB, 20 μg/ml cAD1. We added 200 μg/ml iAD1 to confirm the inhibitory execution of the information gate. OG1RF(pAM714) was then combined with 100-fold excess of OG1SSp and incubated at 37°C for 90 min. OG1SSp was separated from OG1RF(pAM714) by antibiotic selection with 1 mg/ml streptmycin. After washing twice with fresh THB, we incubated OG1SSp for 18 h to increase the total cell number and to esnure that the conjugative plasmids were distributed among the cells.

Confirmation of the Gate Execution

During an 18-h incubation of the final step, we took a sample of the cell mixtures and spread it on an THB agar plate containing 50 μg/ml erythromycin to confirm the result of the gate execution. The total colony number on the plate was counted (Figure 6.4). None of the OG1SSp(pAM714) colony was observed on the following day, but they grew slowly and reached maximum growth after a few days. In the case of OG1SSp(pCF10), the growth delay was not observed. The presence of a high concentration of antibiotic substances creates pressure on the bacteria that even defeats any antibiotic resistance gene. We note that perhaps the condition should be less stringent in the selection step of OG1SSp(pAM714).

A Logical Inverter: Execution of Two Information
Gates Simultaneously

The information gate is applicable to building a logic inverter (Figure 6.5). The gate contains donors with plasmid P_0 and donors with plasmid P_1. Donors with plasmid P_{1-n} ($n = 1$ or 0) in the logic inverter are activated when the inverter receives inhibitor I_n. Activated donors conjugate with recipients and the plasmid P_{1-n} transferred to the recipients. The recipients produce inhibitors I_{1-n} in accordance with the genetic information of the newly obtained plasmid. These then act as the input signal to the next operation.

An execution of signal inversion P_0 to P_1 was tested as follows. We cultured *E. faecalis* strains OG1SSp ($P_0::Tc^r$), OG1RF ($P_0::Tc^r$), OG1RF ($P_1::Em^r$), and OG1SSp at 37°C for 18 h. First, OG1SSp ($P_0::Tc^r$) was used as the input cells which send inhibitors I_0 to the gate. OG1RFs were used as the plasmid

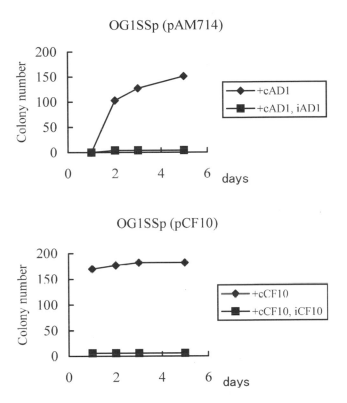

Figure 6.4 Detected colony number.

donors in the gate. OG1SSp without any plasmids was used as the plasmid recipient. The supernatant of OG1RFs was substituted with that of OG1SSp ($P_0::Tc^r$). After incubation at 37°C for 60 min, donors and recipients were combined and incubated for 30 min. We selected the recipients for resistance to streptomycin. It was confirmed that the most of recipients carried P_1 (data not shown). Then, taking the recipients as the next input cells, we also tested the execution of the signal inversion P_1 to P_0.

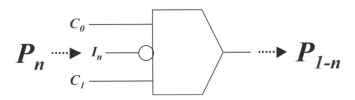

Figure 6.5 A logic inverter based on *E. faecalis* information gate.

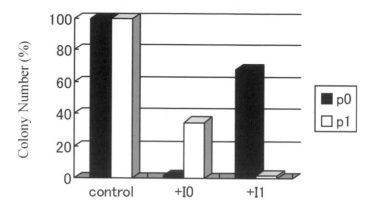

Figure 6.6 Regulation of the plasmid transfer.

Realization of the Logic Inverter

We incubated OG1RF(pCF10), OG1RF(pAM714), and OG1SSp in THB at 37°C for 18 h. After washing twice with fresh THB, OG1RF(pCF10) and OG1RF(pAM714) was combined and incubated for 60 min in THB, 6 μg/ml cCF10, and 20 μg/ml cAD1. We added 200 μg/ml iCF10 and/or 200 μg/ml iAD1 to confirm the inhibitory execution of the information gate. The cell mixtures were then combined with 100-fold excess of OG1SSp and incubated at 37°C for 30 min. OG1SSp was separated from OG1RF by antibiotic selection with 1 mg/ml streptmycin. After washing twice with fresh THB, OG1SSp was incubated for 18 h to increase the total cell number and to ensure that the conjugative plasmids were distributed among the cells.

Confirmation of the Execution of the Logic Inverter

At the final stage we took a sample of cell mixtures and spread it on THB agar plates containing 50 μg/ml erythromycin/1 mg/ml streptmycin or 10 μg/ml tetracycline/1 mg/ml streptmycin to confirm the result of the gate execution. After incubation for 42 h, we counted the total colony number on the each plate (Figure 6.6). The colony number is shown at a scale relative to the control, which was incubated in the presence of 200 μg/ml pheromones and no inhibitor. Each data entry is the average value calculated from three experiments under identical conditions.

CONCLUSIONS

In this work, we showed one way in which bacteria have computing capabilities. We showed how to build a logic circuit with *E. faecalis* by using their communication system. The plasmids carried static 1-bit information. Using the plasmids, we built a logical inverter. To apply the gate to more advanced computing problems, it will be necessary to introduce another technology to rewrite the plasmids. However, little is known about rewriting DNA information with high efficiency in free-living cells.

Previously, various computing models have been proposed based on chemical reactions [7] and DNA [2]. Compared to these models, the reaction speed of our information gate is rather slow. Our information gate may be more suitable for use in applications such as biological sensors than for mainstream computing devices. Nevertheless, this bacterial information gate seems to have some advantages. First, it is plasmid DNA that serves as a medium for exchanging information between individuals. The specific signal, which is composed of pheromones and inhibitors, regulates the system execution. Finally, the signal itself can be written in the plasmid DNA. The plasmid leaves some room to integrate various genes other than the gate signal. Taking these factors into consideration, the information gate seems to be scalable.

Of course, many unsolved problems remain. For example, it is difficult to rewrite the plasmid *in vivo*. The membrane structures of bacteria are generally simple. In the case of *E. faecalis*, all of the cell contents, including the genomic DNA, are located inside a single cell-surface membrane. Rewriting plasmids within the cell involves the unexpected alteration of the genome DNA, usually resulting in cell death. Pheromones released by recipients affect every donor in culture. The recipient may choose the type of donor to conjugate, but cannot choose any individual specifically. Homogenous conjugation between donors may sometimes arise. This type of conjugation merely results in the exchange to plasmids between two donors. It is difficult to transfect the conjugative plasmids into the cells because of their large size. Unlike ordinary plasmid vectors, the conjugative plasmids have few unique restriction sites. Alternative methods such as homologous recombination are required to transform these large-scale plasmids.

The technology of protein engineering is still under development. We have virtually no capability to design artificial proteins with a desired function *de novo*, even less so for artificial cells. All we can do at the moment is to combine artifacts found in nature and perhaps improve them.

Acknowledgments   All of the *E. faecalis* strains and plasmids used in this project were obtained from the H. Nagasawa laboratory. We thank J. Nakayama for his contributions and helpful advice.

References

[1] H. Abelson, D. Allen, D. Coore, C. Hanson, G. Homsy, T. F. Knight, R. Nagpal, E. Rauch, G. J. Sussman, and R. Weiss. Amorphous computing. *Comm. ACM*, 43(5):74–82, 2000.

[2] L. Adleman. Molecular computation of solutions to combinatorial problems. *Science*, 266:1021–1024, 1994.

[3] D. B. Clewell. Bacterial sex pheromone-induced plasmid transfer. *Cell*, 73:9–12, 1993.

[4] D. B. Clewell and K. E. Weaver. Sex pheromones and plasmid transfer in *enterococcus faecalis*. *Plasmid*, 21:175–184, 1989.

[5] G. M. Dunny, B. L. Brown, and D. B. Clewell. Induced cell aggregation and mating in *Streptococcus faecalis*: evidence for a bacterial sex pheromone. *Proc. Natl. Acad. Sci. USA*, 75:3479–3483, 1978.

[6] W. C. Fuqua, S. Winans, and Greenberg E. P. Quorum sensing in bacteria: the luxr-luxi family of cell density-responsive transcriptional regulators. *J. Bacteriol.*, 176:269–275, 1994.

[7] A. Hjelmfelt, E. D. Weinberger, and J. Ross. Chemical implementation of neural networks and turing machines. *Proc. Natl. Acad. Sci. USA*, 88:10983–10987, 1991.

[8] R. Weiss and G. Homsy. Toward in-vivo digital circuits. In L. F. Landweber and E. Winfree, editors, *Evolution as Computation*, pages 275–295. Springer, Berlin, 2003.

[9] R. Weiss and T. F. Knight. Engineered communications for microbial robotics. Proc. 6th International Workshop on DNA-Based Computers, Lecture Notes in Computer Science, vol. 2054, pages 1–16. Springer-Verlag Berlin, Heidelberg, 2001.

[10] R. Wirth. The sex pheromone system of *Enterococcus faecalis*. *Eur. J. Biochem.*, 222:235–246, 1994.

Appendix. Materials

We used *Enterococcus faecalis* strains OG1X, OG1RF, OG1SSp. OG1X has antibioticresistance gene for streptmycin. OG1X is a gene-deletion mutant for the gelatinase that breaks down pheromone peptides. OG1RF has antibiotic resistance for retampicin/fusidic acid. OG1SSp has antibiotic resistance for streptmycin/spectinomycin. *E. faecalis* strains were grown in Todd-Hewitt broth (Difco Laboratories, Detroit, MI). The solid medium used was Todd-Hewitt broth with 1.5% agar. The conjugative plasmid pCF10 contains the antibiotic resistance gene for tetracycline. The conjugative plasmid pAM714, a derivative of pAD1, contains the erythromycin-resistance gene. Antibiotics were used at the following concentrations: tetracycline, 10 μg/ml; erythromycin, 20 μg/ml; streptomycin, 1 mg/ml; fusidic acid, 50 μg/ml. Synthetic peptides cCF10, cAD1, iCF10, and iAD1 were prepared at NikkaTechno service.

7

Cellular Computation and Communication Using Engineered Genetic Regulatory Networks

Ron Weiss, Thomas F. Knight Jr.,
and Gerald Sussman

In this chapter we demonstrate the feasibility of digital computation in cells by building several operational *in vivo* digital logic circuits, each composed of three gates that have been optimized by genetic process engineering. We have built and characterized an initial cellular gate library with biochemical gates that implement the NOT, IMPLIES, and AND logic functions in *E. coli* cells. The logic gates perform computation using DNA-binding proteins, small molecules that interact with these proteins, and segments of DNA that regulate the expression of the proteins. We also demonstrate engineered intercellular communications with programmed enzymatic activity and chemical diffusions to carry messages, using DNA from the *Vibrio fischeri lux* operon. The programmed communications is essential for obtaining coordinated behavior from cell aggregates.

This chapter is structured as follows: the first section describes experimental measurements of the device physics of *in vivo* logic gates, as well as genetic process engineering to modify gates until they have the desired behavior. The second section presents experimental results of programmed intercellular communications, including time–response measurements and sensitivity to variations in message concentrations.

MEASURING AND MODIFYING DEVICE PHYSICS

Potentially the most important element of biocircuit design is matching gate characteristics. Experimental results in this section demonstrate that circuits with mismatched gates are likely to malfunction. In generating biology's complex genetic regulatory networks, natural forces of selection have resulted in finely tuned interconnections between the different regulatory components. Nature has optimized and matched the kinetic characteristics of these elements so that they cooperatively achieve the desired regulatory behavior. In building *de novo* biocircuits, we frequently combine regulatory elements that do not interact in their wild-type settings. Therefore, naive coupling of these elements will likely produce systems that do not have the desired behavior.

In *genetic process engineering*, the biocircuit designer first determines the behavioral characteristics of the regulatory components and then modifies the elements until the desired behavior is attained. Below, we show experimental results of using this process to convert a nonfunctional circuit with mismatched gates into a circuit that achieves the correct response. The experiments focus on examining and modifying the steady-state behavior of the genetic circuits and represent the first example of designing robust genetic regulatory components for use in building reliable biocircuits of significant complexity. Future work will also consider the dynamic behavior of the circuits. The research reported in this section represents the beginning of a process to assemble a library of components with known and useful device physics, akin to the TTL Data Book for electrical circuit design. The knowledge of device physics plays a fundamental role in achieving predictable and reliable biocircuit design.

Figure 7.1 shows the wiring diagram of genetic circuits we constructed to measure the device physics of two seperate inverters, one based on the *lacI* repressor/p(lac) promoter, and the other based on the *cI* repressor/$\lambda_{P(R)}$ promoter. Because the R_2 repressor input to the P_2 IMPLIES ([NOT x] OR y)

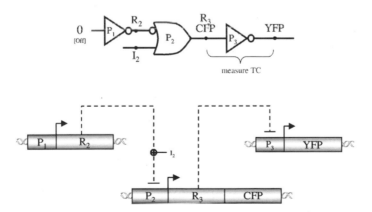

Figure 7.1 Genetic circuit diagram to measure the device physics of an R_3/P_3 inverter: digital logic circuit and the genetic regulatory network implementation (P_x: promoters, R_x: repressors, CFP/YFP: reporters).

gate is constantly high, the level of the inducer molecule input I_2 determines the level of the repressor R_3. R_3 is the input protein to the R_3/P_3 inverter gate under study. The cyan fluorescent protein (CFP) transcribed along with R_3 reports the level of the input signal, while the yellow fluorescent protein (YFP) simultaneously reports the output signal expressed from the R_3/P_3 inverter. The output of this circuit is the logical NOT of the inducer input signal.

A transfer function is the relation between the input signal and the output signal of a gate or a circuit in steady state. Analog ranges represent digital signals of zeros and ones. An ideal transfer curve for an inverter has an inverse sigmoidal shape: the gain (or slope) is flat, then steep, then flat again. Because of the gain, the output of the inverter is a better representation of the digital value then the input (i.e., signal restoration). Figure 7.2 shows transfer curves of three circuits with an inverter based on $cI/\lambda_{P(R-O12)}$. The flat curve represents the nonresponsive behavior of a circuit with an inverter based on the original $cI/\lambda_{P(R-O12)}$ genetic elements. This demonstrates that coupling genetic components into a circuit without first understanding their device physics may yield completely nonfunctional systems. Later in this section we describe genetic mutations performed on the original $cI/\lambda_{P(R-O12)}$ genetic elements to obtain a functional circuit with the desired input/output behavior. Figure 7.2 shows how two mutations result in an inverse sigmoidal transfer curve with good gain and noise margins.

External Control of Signals

The first step in measuring the device physics of an inverter is to construct genetic circuits that allow the researcher to externally set the *in vivo* level of a signal. This is performed using circuits where an inverter is connected to an

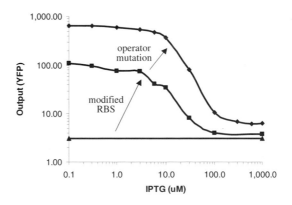

Figure 7.2 Genetic process engineering of the $cI\lambda_{P(R-O12)}$ inverter. A series of genetic modifications converts a nonfunctional circuit into one that achieves the desired input/output behavior.

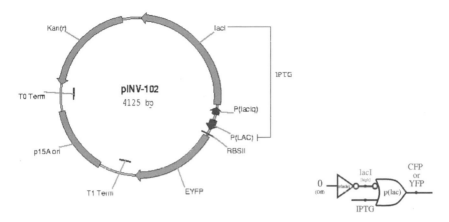

Figure 7.3 Genetic circuit to set protein expression levels. Isopropylthio-β-galacto-side (IPTG) concentration controls the level of the output protein, enhanced yellow fluorescent protein.

IMPLIES gate. We constructed two such circuits, one on plasmid pINV-102 with the enhanced yellow fluorescent protein (EYFP) output protein (Figure 7.3) and another circuit on a similar plasmid named pINV-112-R1 with the enhanced cyan fluorescent protein (ECFP) as the output protein. The fluorescent proteins are both from Clontech [7]. The inverter that comprises the constitutive promoter p(lacIq) has an input that is always set to low because the cell does not contain a repressor for the p(lacIq) promoter. Therefore, the output of the inverter, *lacI*, is constantly high. Then, since the *lacI* repressor input to the p(lac) IMPLIES gate is constantly high, the level of the inducer molecule input, IPTG (isopropylthio-β-galactoside), is positively correlated with the level of the output. The researcher controls the level of the output signal with this circuit by externally setting the level of IPTG, which freely diffuses into the cell.

In Figure 7.3, the double-stranded plasmid layout includes the following genetic elements: the p15A origin controlling the copy number of the plasmid in the cell, kanamycin antibiotic resistance for selective growth, promoters represented by short arrows, protein-coding sequences downstream of specific promoters, and transcription terminators such as T1 term. The promoters and protein-coding sequences are either clockwise or counterclockwise depending on the direction of the DNA strand that encodes them.

Figure 7.4 shows data from *Escherichia coli* cells with the pINV-102 and pINV-112-R1 plasmids grown for approximately 5 h in culture until they reached steady state. The construction of all plasmids and experimental conditions in this chapter are described in detail elsewhere [15]. The data include median fluorescence values obtained using fluorescence-activated cell sorting (FACS) [12] of the different cell populations induced with a range of IPTG

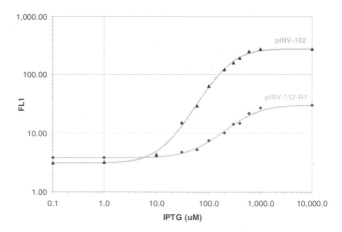

Figure 7.4 Controlling signal levels using external induction with isopropylthio-β-galactoside (IPTG).

concentrations. The graph shows how to control an *in vivo* signal using external induction with IPTG. The relationship between the ECFP and EYFP fluorescence intensities in Figure 7.4 is used to normalize between simultaneous ECFP/EYFP readings in subsequent experiments in this chapter. This genetic setup is used in the following sections to set the levels of input mRNA to the inverters under study.

The lacI/p(lac) Inverter

Figure 7.5 shows the genetic circuit used to measure the device physics of an inverter based on the *lacI* repressor and the p(lac) promoter. The first two logic gates set the level of the input signal to the inverter in a mechanism similar to one used in the circuit described in the previous subsection. Here, the $\lambda_{P(R-O12)}$ inverter functions as a constitutive promoter (no *cI* in the system) to set a constant high level of the Tet repressor (*tetR*). Then, through the *tetR*/P(LtetO-1) IMPLIES gate, the concentration of the aTc (anhydrotetracycline) inducer molecule controls the level of the lac repressor (*lacI*). *lacI* is the input protein to the inverter gate under study. The ECFP transcribed along with *lacI* reports the level of the input signal. Finally, EYFP reports the output signal expressed from the lacI/p(lac) inverter.

   The output of this circuit is the logic NOT of the aTc input signal. Figure 7.6a shows FACS cell population data of the EYFP output signal in seperate experiments where the cells were exposed to different aTc inducer input concentrations. For a low input concentration of 3 ng/ml aTc, the output of the circuit is appropriately high. For a high input concentration of 30 ng/ml aTc, the output of the circuit is correctly low. The figure also illustrates the good

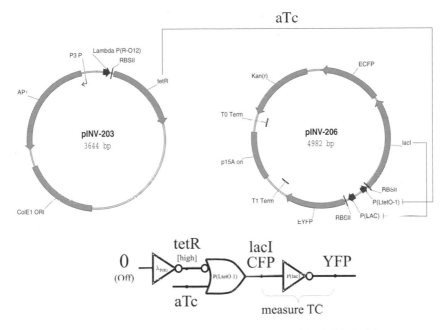

Figure 7.5 Genetic circuit to measure the transfer curve of the lacI/p(lac) inverter.

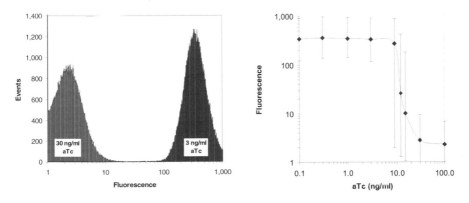

(a) Digital computation noise margins   (b) Gain/signal restoration

Figure 7.6 Transfer curve gain and noise margins for the lacI/p(lac) circuit.

noise margins between low and high signal values because the two aTc input levels shown are immediately before and after the sharp transition from high to low output.

Figure 7.6b illustrates the transfer function of the circuit with respect to the level of the inducer. The figure relates aTc input levels to YFP output fluorescence levels, with error bars depicting the range that includes 95% of the flow cytometry fluorescence intensities of the cells recorded for the particular aTc level. The favorable noise margins and signal restoration of this circuit clearly demonstrate that digital-logic computation is feasible with genetic circuits.

By correlating ECFP and EYFP readings for the same experiment, Figure 7.7 shows the normalized transfer curve of the lacI/p(lac) inverter. The ECFP fluorescence intensities are normalized to the EYFP levels based on the experimental results described in the previous subsection. After normalization, each point represents the median fluorescence intensities for a particular experimental condition of the input signal (ECFP) versus the output signal (EYFP). The gain of the inverter of 4.72 is sufficient for digital-logic computation and is likely to be related to the pentameric nature of *lacI* repression [9].

The lacI/p(lac) gate characterized in this section is the first component of the cellular gate library. Next, we describe the addition of the second component to the gate library, the $cI/\lambda_{P(R-O12)}$ inverter. In particular, the following section demonstrates genetic process engineering to optimize the original behavior of this gate and to obtain the desired behavior for digital computation.

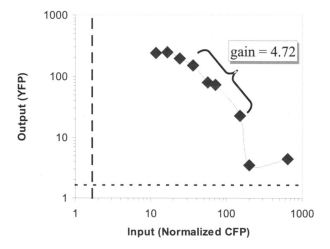

Figure 7.7 The lacI/p(lac) transfer curve.

Optimizing $cI/\lambda_{P(R-O12)}$ Inverters

Although the gain exhibited by the lacI/p(lac) is sufficient, other repressor/promoter combinations can yield better signal restoration. One such example is the $cI/\lambda_{P(R-O12)}$ inverter. The cI repressor binds cooperatively to the λ promoter's O_R1 and O_R2 operators, which results in high gain. The $\lambda_{P(R-O12)}$ is a synthetic promoter that we designed and that excludes O_R3 of the wild-type λ promoter. The affinity of cI to O_R3 is weak, and this operator would not significantly enhance the repression efficiency of cI.

The cI monomer, also known as the λ repressor, has an amino domain composed of amino acids 1–92, a carboxyl domain of residues 132–236, and 40 remaining amino acids that connect the two domains [10]. The monomers associate to form dimers, which can then bind to the 17 basepair operator regions. The intrinsic affinity of cI to O_R1 is about 10 times higher than to O_R2, and therefore typically cI binds O_R1 first. However, the binding of cI to O_R1 immediately increases the affinity of a second dimer to O_R2 because of the interaction with the previously bound dimer. As a result, repressor dimers bind to O_R1 and O_R2 almost simultaneously. From a circuit engineering perspective, this cooperative binding leads to a much desired high gain because the transition from low repression activity to high repression activity occurs over a small range of repressor concentrations.

The genetic circuit to measure the device physics of the $cI/\lambda_{P(R-O12)}$ is logically similar to the one used to measure the $lacI$/p(lac) inverter (Figure 7.8). Here, the $lacI$/p(lac) IMPLIES gate controls the level of the cI input. For a particular experiment, the researcher sets the repressor level by controlling the IPTG concentration.

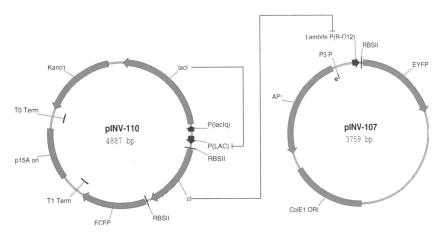

Figure 7.8 Genetic circuit to measure the transfer curve of $cI/\lambda_{P(R-O12)}$, using $lacI$ as driver.

The logic interconnect of this circuit should result in EYFP fluorescence intensities that are inversely correlated with the IPTG input levels. However, as shown in Figure 7.2 (the flat line), the circuit is completely unresponsive to variations in the IPTG levels. The lack of response stems from the mismatch between the kinetic characteristics of the *lacI*/p(lac) gate versus the *cI*/$\lambda_{P(R-O12)}$ inverter. Specifically, with no IPTG, the fully repressed expression level from p(lac) still results in a low level of *cI* mRNA. Because the ribosome binding site is very efficient, the low mRNA level results in some translation of the *cI* protein. And because *cI* is a highly efficient repressor, even a low concentration represses the $\lambda_{P(R-O12)}$ promoter to the point where no fluorescence can be detected. This gate mismatch highlights the importance of understanding the device physics of the cellular gates. The following two subsections describe genetic process engineering to modify genetic elements in the *cI*/$\lambda_{P(R-O12)}$ inverter such that the gate obtains the desired behavioral characteristics.

Modifying Ribosome-Binding Sites

Ribosome-binding site (RBS) sequences significantly control the rate of translation from the input mRNA signal to the input protein (Figure 7.9). These sequences align the ribosome onto the mRNA in the proper reading frame so that polypeptide synthesis can start correctly at the AUG initiation codon. The affinity of the ribosome's 30S subunit to the RBS that it binds determines the rate of translation. This translation rate is included in a biochemical reaction for modeling and simulating the inverter using BioSPICE [15, 16]. For a given input mRNA level, a reduction in the translation rate yields a lower input protein

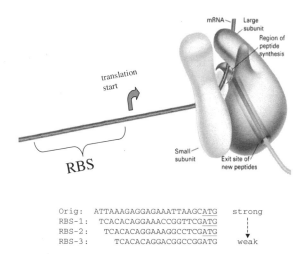

```
Orig:   ATTAAAGAGGAGAAATTAAGCATG     strong
RBS-1:  TCACACAGGAAACCGGTTCGATG
RBS-2:  TCACACAGGAAAGGCCTCGATG
RBS-3:  TCACACAGGACGGCCGGATG        weak
```

Figure 7.9 The ribosome, mRNA, and different ribosome-binding sites.

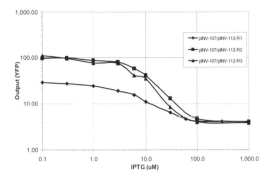

Figure 7.10 The effect of weaker ribosome binding sites on the behavior of the $cI/\lambda_{P(R-O12)}$ circuit. Here the output, YFP, exhibits the desired inverse sigmoidal relationship to the isopropylthio-β-galactoside (IPTG) input.

level, which pushes the entire transfer curve upward and outward. BioSPICE simulations serve as an abstract model to study the effects of changing reaction kinetics of specific genetic components in the logic computation.

The sequences for the original highly efficient cI RBS used in the circuit above, as well as three other less efficient RBS's from the literature [6] are:

orig:	ATTAAAGAGGAGAAATTAAGC<u>ATG</u>	(strongest)
RBS-1:	TCACACAGGAAACCCGTTCG<u>ATG</u>	
RBS-2:	TCACACAGGAAAGGCCTCG<u>ATG</u>	
RBS-3:	TCACACAGGACGGCCGG<u>ATG</u>	(weakest)

Starting from pINV-110, we constructed three new plasmids (pINV-112-R1, pINV-112-R2, pINV-112-R3) where the three weaker RBS's replace the original RBS of the cI. pINV-112-R1 contains the strongest RBS, while pINV-112-R3 contains the weakest RBS.

Figure 7.10 shows the dramatic effect of the RBS change on the behavior of the circuit, where now the output YFP exhibits the desired inverse sigmoidal relationship to the IPTG input. The circuit with the strongest RBS (pINV-107/pINV-112-R1) shows a moderate sensitivity to IPTG, while the circuits with the other two RBSs display a more pronounced response to variations in IPTG.

Modifying Repressor/Operator Affinity

Replacing the strong RBS with weaker sites converted a nonfunctional circuit into a functional one and demonstrated the utility of genetic process engineering. This subsection describes further modifications to the repressor/operator affinity that yield additional improvements in the performance of the circuit. These modifications are also motivated by BioSPICE simulations [15, 16]. The simulations show how reductions in the binding affinity of the repressor to the operator reshape the transfer curve of the inverter upward and outward.

To reduce the repressor/operator affinity, we constructed three new plasmids with modified O_R1 sequences using site-directed mutagenesis to have the following sequences:

orig:	TACCTCTGGCGGCGGTGATA
mut4:	TACATCTGGCGGCGGTGATA
mut5:	TACATATGGCGGCGGTGATA
mut6:	TACAGATGGCGGCGGTGATA

cI's amino domain is folded into five successive stretches of α helix, where α helix-3 lies exposed along the surface of the molecule [10]. This α helix recognizes the λ operators and binds the repressor to those particular DNA sequences. The two α helix-3 motifs of the repressor's dimer complex are separated by the same distance as the one separating successive segments of the major groove along one face of the DNA. These motifs efficiently bind the repressor dimer to the mostly symmetric λ operator regions, where each operator consists of two half-sites. The following is the consensus sequence for the 12 operator half-sites in the wild-type bacteriopage λ (subscripts correspond to the frequency of the basepair in the given position):

T_9	A_{12}	T_6	C_{12}	A_9	C_{11}	C_7	G_9	C_5
C_2		C_3		T_2	T_1	T_4	T_2	T_1
A_1		A_1		C_1		G_1	C_1	

In choosing mutations to perform, we conjectured that bases with high frequency in the consensus sequence would be significant to strong repressor/operator binding. Mut4 is a one-basepair mutation $C \rightarrow A$ of the fourth O_R1 position; mut5 is a two-basepair mutation that also modifies the sixth O_R1 position $C \rightarrow A$; and mut6 is a three-basepair mutation that also modifies the fifth O_R1 position $T \rightarrow G$.

The experimental results in Figure 7.11 demonstrate the effect of coupling the three $\lambda_{P(R-O12)}$ O_R1 operator mutations with the weakest RBS from above. The two- and three-basepair O_R1 mutations, coupled with the weak RBS, produce a circuit where the highest levels of cI cannot repress the output of the $cI/\lambda_{P(R-O12)}$ gate. A one-basepair mutation to O_R1 in plamids pINV-107-mut4/pINV-112-R3 yields a circuit with a well-behaved response to the IPTG signal and is a good gate candidate for other biocircuits.

As described in this section, using genetic process engineering we first examined the behavioral characteristics of the $cI/\lambda_{P(R-O12)}$ inverter and then genetically modified the gate until we produced a version with the desired inverse sigmoidal behavior. The design and experimental results illustrate how

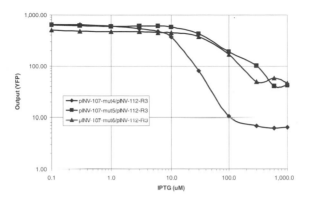

Figure 7.11 The effect of $\lambda_{P(R-O12)}$ O_R1 operator mutations on the behavior of the $cI/\lambda_{P(R-O12)}$ circuit.

to convert a nonfunctional circuit with mismatched gates into a circuit that achieves the correct response (Figure 7.2). Note that in choosing genetic variations, the specific RBS modifications described earlier can be applied to any inverter and are independent from the specific genetic candidate. The operator mutations in this subsection are specific to the $\lambda_{P(R)}$ and cI and cannot be generally applied to other operators. Accordingly, there are several strategies to optimizing a new component. First, one can test modifications that are applicable to any component. Second, one can study the particular component by reading the literature and performing laboratory experiments and choosing mutations based on the understanding of the specific biochemistry of the element. Third, one can perform large scale, random mutations on the element and screen for mutants that have the desired behavior.

The device physics measurements in this section facilitate biocircuit design because they enable prediction of the behavior of complex circuits using the characteristics of simple components. With the appropriate tools, the engineer of biocircuits can begin to design and produce large-scale circuits. In understanding the device physics of cellular gates, great emphasis must be placed on considering the special substrate properties, such as signal fluctuations and matching kinetic characteristics. The experimental results reported here represent an important effort to assemble a library of simple standardized biological components with the proper device physics that can be combined in predictable ways to engineer novel cell behaviors.

CELL–CELL COMMUNICATIONS

In this section, we describe engineered cell–cell communications using chemical diffusion. Communication between cells is obviously essential to any kind of

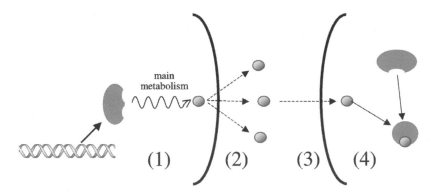

Figure 7.12 Cell-to-cell communication schematics: (1) The sender cell produces small signal molecules using certain metabolic pathways. (2) The small molecules diffuse outside the membrane and into the environment. (3) The signals then diffuse into neighboring cells and (4) interact with proteins in the receiver cells, thereby changing signal values.

coordinated behavior. However, in cell aggregates óne kind of communication emerges as especially important—the ability to detect and act on chemical signal concentration gradients. Such gradient-dependent expression is the building block of locally unique behavior and cell differentiation, providing one basic organizing principle for complex patterned development.

We have isolated a specific chemical cell-to-cell signaling mechanism from a natural biological system, the quorum sensing system of *Vibrio fischeri*. This system encodes genes and promoter sequences that allow the controlled expression of the chemical *Vibrio fischeri* autoinducer (VAI) within one sender cell, and the detection and controlled expression of specific genes in another receiving cell. The free diffusion of the VAI chemical within the medium and across cell membranes allows the establishment of chemical gradients and the controlled expression of genetic circuits as a result.

Specifically, we describe the construction and testing of engineered genetic circuits that exhibit the ability to send a controlled signal from one cell, diffuse that signal through the intercellular medium, receive that signal within a second cell, and activate a remote transcriptional response (Figure 7.12). The work reported in this section implements and characterizes cell-to-cell communication components for the cellular gate library.

Quorum Sensing in Bacteria

Vibrio fischeri is a Gram-negative, bioluminescent, marine prokaryote that naturally occurs in two distinct environments. In seawater, it swims freely at concentrations of approximately 10 cells/l. It also grows naturally in a symbiotic relationship with a variety of invertebrate and vertebrate sea organisms, especially the Hawaiian sepiolid squid, *Euprymna scolopes*, and the Japanese pinecone fish, *Monocentris japonica* [11]. In these symbiotic relationships, the bacteria grows to densities of approximately 10^{10} cells/l.

In the free-living state, *Vibrio fischeri* emits essentially no light (< 0.8 photons/sec/cell). In the light organ of the Hawaiian sepiolid squid, however, the same bacteria emit more than 800 photons/sec/cell, producing very visible bioluminescence. In culture, *Vibrio fischeri* demonstrates a similar density-dependent bioluminescence, with induction occurring at about 10^{10} cells/l.

Work over many years has established that this behavioral change is due to a natural mechanism of detecting cell densities, which has been termed *quorum sensing* [5]. The quorum-sensing mechanism relies on the synthesis and detection of a very specific, species-unique chemical, an *autoinducer*, which mediates intercellular communications. In *Vibrio fischeri*, this autoinducer chemical (VAI) has been identified as *N*-(3-oxohexanoyl)-3-amino-dihydro-2-(3H)-furanone [2]. The gene, *LuxI*, catalytic protein, and synthetic pathway for this chemical have also been identified [4].

The *LuxI* gene encodes an acyl-homoserine lactone synthesase that uses highly available metabolic precursors found within most Gram-negative prokaryotic bacteria—acyl-ACP from the fatty acid metabolic cycle, and *S*-adenosylmethionine (SAM) from the methionine pathway—to synthesize VAI. The VAI freely diffuses across the bacterial cell membrane. Thus, at low cell densities, low VAI concentrations are available. Within a light organ, or at high culture densities, VAI builds up within the environment, resulting in a density-dependent induction of bioluminescence.

The response mechanism to VAI concentration has also been extensively analyzed [13]. The *LuxR* gene codes for a two-domain DNA-binding protein that interacts with VAI and the Lux box of the LuxICDABEG operon promoter to exercise transcriptional control (Figure 7.13). At nanomolar concentrations, VAI binds to the N-terminal domain of the LuxR protein, which in turn activates

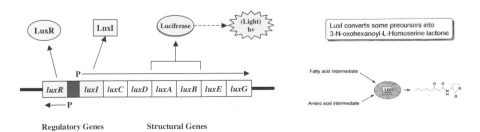

Figure 7.13 The lux operon (on the left), and luxI metabolism that catalyzes the formation of *V. fischeri autoinducer* (on the right).

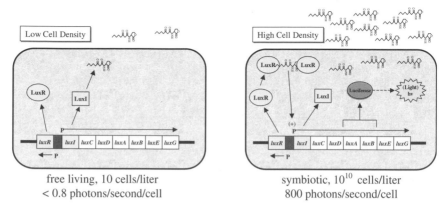

Figure 7.14 The lux operon and quorum sensing: density-dependent biolumin-escence.

the C-terminal helix-turn-helix DNA-binding domain. The LuxR protein acts as a transcriptional activator for the RNA polymerase holoenzyme complex. The activated protein likely binds in dimeric or multimeric forms because of the evident dyadic symmetry of the Lux box binding domain. Figure 7.14 shows the genetics of bioluminescence in two distinct environments in which *Vibrio fischeri* naturally occurs. The dimeric binding of the LuxR product to the operator produces the kind of nonlinear concentration/response behavior [8, 16] and widely seen in DNA-binding protein transcriptional control. This nonlinear response is an essential element of signal restoration and digital control of expression.

Successful cloning and expression of the *Lux* genes into *E. coli* have established the genetic structure of the *Vibrio fischeri* Lux operon [3]. It is somewhat surprising (although common) for the transfer of regulatory genes and entire metabolic pathways to function straightforwardly across Gram-negative species boundaries in this way.

Intercellular Signaling Experiments

To experiment with engineered cell-to-cell communications, we constructed a series of plasmids, as described elsewhere [15]. The plasmids encode genetic logic circuits that enable cells to send messages and logic circuits that enable cells to detect and respond to incoming messages. This section describes experiments to characterize the cell-to-cell communication capabilities engineered into *E. coli* cells using these genetic circuits.

Sending a Constant Cell-to-Cell Signal

The first intercellular communications experiment involved sending a constant signal from one cell type to another. The pSND-1 plasmid encodes a circuit that

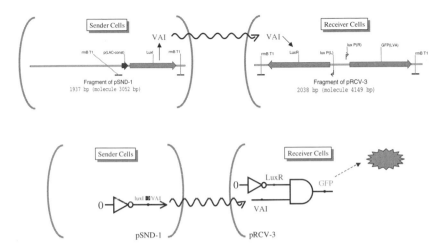

Figure 7.15 Genetic and logic circuits for pSND-1 sender and pRCV-3 receiver. The sender cells constitutively express *luxI*, which catalyzes the formation of *V. fischeri autoinducer* (VAI). VAI diffuses into the environment and neighboring cells, which detect VAI through the transcriptional activation of lux P(R).

directs the cell to continuously send the VAI message. The pRCV-3 plasmid encodes a circuit that directs the cell to express GFP(LVA), a variant of the green fluorescent protein from Clontech, when VAI enters the cell (Figure 7.15). Cultures of *E. coli* DH5α transformed with the pRCV-3 plasmid and cultures of *E. coli* DH5α transformed with the pSND-1 plasmid were grown separately overnight at 37°C in Luria-Bertani ampicillin (LB AMP). A 96-well clear-bottom plate was loaded with 200 μl of LB AMP in each well. We loaded 10 μl of pSND-1 cells horizontally to each well, along with controls consisting of cells expressing GFP(LVA) constitutively with the pRW-LPR-2 plasmid, *E. coli* DH5α containing pUC19 to serve as a negative control, and a series of wells containing extracted VAI (see below).

Vertically, 10 μl of cells containing the pRCV-3 construct were also loaded into each well. Thus, each well contained a variety of senders and a uniform set of receivers. The plate was grown in a Biotek FL-600 fluorescent plate reader for 2 h, and fluorescence at the GFP(LVA) peak (excitation filter 485/20 nm, emission filter 516/20 nm) was measured every 2 min. Figure 7.17 shows the time-series response of the different cultures. Wells containing only the pRCV-3 cells, or with added pUC19 cells, showed no increase in fluorescence. The well containing pRCV-3 cells and pRW-LPR-2 cells (which express GFP[LVA]) served as a positive control for high levels of fluorescence. Wells containing the pRCV-3 cells plus extracted pTK1 autoinducer showed high and increasing levels of fluorescence. Cells with pRCV-3 and pSND-1 showed the expected increase in fluorescence demonstrating successful cell-to-cell signaling.

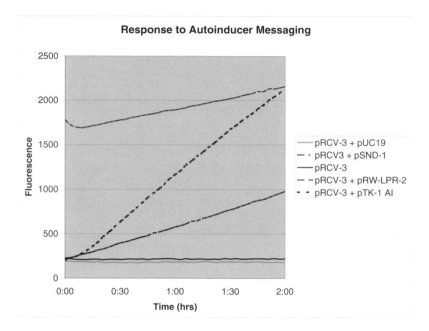

Figure 7.16 Verification of communication constructs.

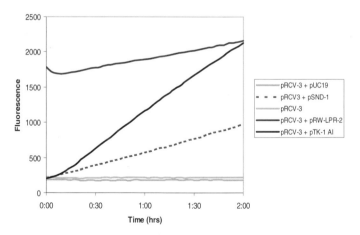

Figure 7.17 Time-series response of receivers to signal. The negative controls (pRCV-3 and pRCV-3) show no increase in fluorescence over time. The positive control with constitutive expression of GFP(LVA) (pRW-LPR-2) shows a high level of fluorescence. The mixed culture of senders and receivers (pRCV-3 + pSND-1) shows the increase in fluorescence over time due to the cell-to-cell communications. Finally, a highly concentrated extract of VAI mixed with the receivers (pRCV-3 + pTK-1 AI) shows a rapid increase in fluorescence.

Characterization of the Receiver Module

The genetic circuit to receive messages (plasmid pRCV-3) was further characterized by inducing the promoter with different levels of VAI extracted from cell culture. Cultures of *Vibrio fischeri* and of *E. coli* containing the pTK1 plasmid were grown overnight to stationary phase in Good Vibrio Medium (GVM) broth or LB Amp, respectively, at 30°C, which allows evaluation of their bioluminescence. After verification of light production, 100 ml of the cultures were centrifuged at 3300*g*, and the supernatant collected. The supernatant was extracted with 10 ml of ethyl acetate by vigorous shaking in a separatory funnel for 10 min. The ethyl acetate extract (upper fraction) was separated and dried under vacuum. The resulting crude extract was redissolved in 1 ml of deionized water to provide 100× VAI extract.

We analyzed the effectiveness of serial dilutions of the VAI extracts from pTK1 and *Vibrio fischeri* in inducing GFP expression of the pRCV-3 cells. Both the *Vibrio fischeri* and pTK1 extracts were about equally effective at inducing expression of the pRCV-3 promoter, as measured by GFP production. Cells with different levels of VAI were incubated at 37°C for 4 h, and the maximum fluorescence achieved for each culture was recorded. Figure 7.18 shows that increasing levels of autoinducer yielded increasing GFP expression by the receiver. High levels of the extract, however, were toxic to the cells and resulted in relatively low fluorescence levels.

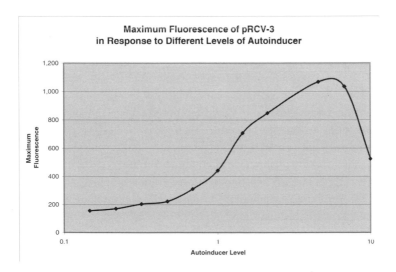

Figure 7.18 The effect of different autoinducer levels on the maximum fluorescence attained by the receivers.

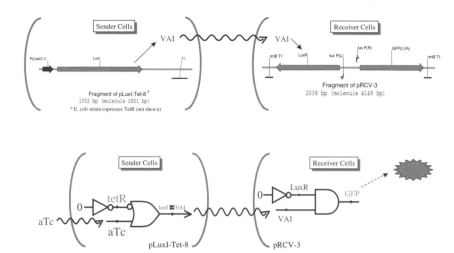

Figure 7.19 Genetic and logic circuits for pLuxI-Tet-8 sender and pRCV-3 receiver.

Sending Controlled Cell-to-Cell Signals
The third experiment characterized the response of the receivers to variations in the strength of the message transmitted by the senders. For this, the *LuxI* gene was placed under the control of the Tet promoter from the Clontech pPROTet system. Figure 7.19 provides a schematic representation of the experiment. In one cell, the pLuxI-Tet-8 plasmid exerts controlled expression of the LuxI autoinducer synthesase using the Tet operon. The synthesase catalyzes the conversion of normal cellular metabolic products into VAI; thus, controlling the *LuxI* expression level controls the VAI production in the cells. The VAI produced within the cells migrates though the cell membrane of the sender, into the culture medium, and through the membrane of the receiver—a cell containing the pRCV-3 plasmid. There, it interacts with the N-terminal domain of the LuxR DNA-binding protein product, disabling it from binding to the Lux box binding site. The expression of the GFP reporter gene is enhanced, resulting in high levels of fluorescence (Figure 7.20).

The experiment involved the incubation of similar mixed-cell cultures on 96-well, clear-bottom plates. One important difference was the culture medium: the pPROTET cells carry spectinomycin and chloramphenicol resistance, while the pRCV-3 cells carry ampicillin resistance. The experiments were carried out by growing overnight cultures of both types of cells in the appropriate antibiotic containing medium, followed by centrifugation at 4000*g* to remove the medium and resuspension to similar cell density in LB containing no antibiotics, so that both cell types could grow.

Three rows of pRCV-3 cells were loaded on a microplate, and two of these rows were also loaded with pLuxI-Tet-8 cells. The sender cells in the var-

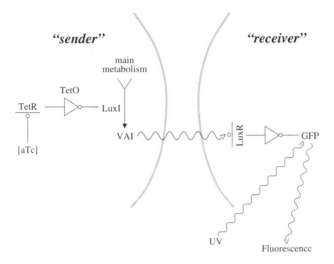

Figure 7.20 Circuit diagram of gradient communications.

ious columns were induced with different levels of aTc. Also, one control column included receivers that were induced directly with the VAI extract. Figure 7.21 shows the results of this experiment after culturing the plate for 4 h at 37°C. As expected, the null wells containing no aTc or VAI showed no enhancement of fluorescence, while the positive control wells with the 10× VAI extract exhibited fluorescence. The experiments labeled BL21-LuxITet include senders where the pLuxl-Tet-8 plasmid was transformed into BL21-PRO cells, while the experiments labeled DH5α-LuxITet include senders where the pLuxl-Tet-8 plasmid was transformed into *E. coli* DH5α cells.[1] In wells containing sender cells induced with aTc at levels lower than about 20 ng/ml, the receiver cells exhibit only only a small fluorescent response. In wells induced with aTc levels great than 200 ng/ml, the receiver cells exhibit a significant response. Sufficiently high levels of aTc inhibited cell growth.

Visual Observation of Communications
Finally, this section describes three visual observation experiments of cell-to-cell communications to understand the diffusion and reaction characteristics. First, pINV-112-R3 and pSND-1 were transformed into *E. coli* JM2.300 cells to create sender cells. With IPTG induction, these sender cells emit cyan fluorescence that serves as an easily identifiable marker due to the ECFP encoded downstream of p(lac) on pINV-112-R3. In addition, the pSND-1 plasmid directs

---

[1] In BL21-PRO cells, TetR (needed for controlled induction of the Tet promoter) exists on a plasmid, while in DH5α TetR is part of the chromosomal DNA.

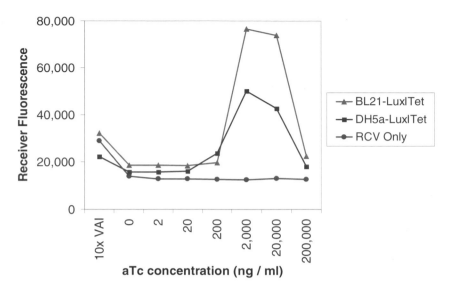

Figure 7.21 Controlling the sender's signal strength: maximum response of receivers to different anhydrotetracycline (aTc) induction of senders. The 10× *V. fischeri* auto-inducer (VAI) positive control column shows the results of introducing the VAI extract into the wells. The rest of the columns illustrate the response of the receivers from induction of the senders with various levels of aTc.

the cells to constitutively express *luxI*, resulting in constant transmission of the message.

Second, pRCV-3 and pPROLar.A122 were transformed into *E. coli* JM2.300 cells to create receiver cells. The pRCV-3 plasmid instructs the cell to express GFP(LVA) when VAI enters the cytoplasm. The pPROLar.A122 plasmid confers kanamycin resistance to the cell. Therefore, these sender and receiver cells have both kanamycin and ampicillin resistance.

In the experiments reported here we used a Nikon Eclipse E800 fluorescence microscope equipped with a Hamamatsu C4742 ORCA I CCD camera controlled by QED Imaging software. The cyan fluorescence filter is chroma cyan GFP (excitation: 436/20 nm, emission: 480/40 nm), the green fluorescence filter is a chroma fluoroscein isothiocyanate/EGFP (excitation: 480/40 nm, emission: 535/50 nm), and the yellow fluorescence filter is a chroma yellow GFP (excitation: 500/20 nm, emission: 535/30 nm).

In the first visual observation experiment, sender and receiver cells were grown separately overnight at 37°C, shaking at 250 rpm, each in 2 ml LB Amp/kanamyain (Kan) and 1 mM IPTG inside 14 ml Falcon polystyrene tubes (352051). Then, the JM2.300[pSND-1/pINV-112-R3] cells were pelleted at 6800 rpm and resuspended in fresh 50 μl LB Amp/Kan 1 mM IPTG. The JM2.300[pRCV-3/pPROLar.A122] cells were pelleted at 6800 rpm and

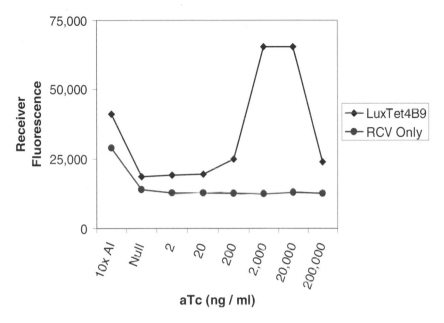

Figure 7.22 Controlling the sender's signal strength: maximum response of receivers to different anhydrotetracycline (aTc) induction of senders.

resuspended in fresh 400 µl LB Amp/Kan and 1 mM IPTG. We spotted 10 µl droplets of receiver cells on an LB Amp/Kan and 1 mM IPTG agar plate and dried them for 10 min. Then, 0.2-µl droplets of sender cells were spotted next to the receivers such that the senders partially overlapped the receiver cells. A quick check under the microscope showed that the sender cells were emitting cyan fluorescence, while the receivers exhibited no fluorescence. The plate was then incubated for 1 h at 37°C.

Figure 7.23 shows the fluorescence pattern of a sender droplet partially overlapping a receiver droplet after the incubation period. The images were captured with a Plan Fluor 4× objective, using the cyan and yellow filters (the yellow filter is better than the green filter for distinguishing between cyan fluorescence and green fluorescence). The physical width of each image is approximately 1.7 mm. The senders were still emitting cyan fluorescence, while the receiver cells were emitting green fluorescence due to the VAI that diffused from the senders. The interesting patterns resulted from liquid diffusion and mixing.

Figure 7.24 shows fluorescence images of the same communication experiment under a higher magnification (40× objective). Each picture represents a montage of three overlapping image captures, with a total physical width of approximately 0.27 mm. At this degree of magnification, some individual bacterial cells can be distinguished.

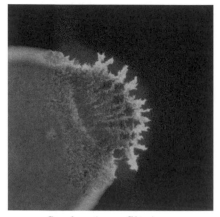

Senders (cyan filter)

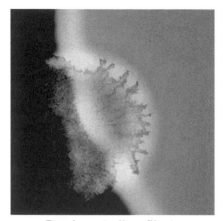

Receivers (yellow filter)

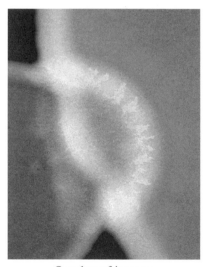

Overlay of images

Figure 7.23 Microscope fluorescence images of overlapping communicating cells (4× objective).

The second visual observation experiment captured a time-lapse series of images that illustrate the dynamic behavior of the communications (Figure 7.25). Sender and receiver cells were grown separately overnight at 37°C, shaking at 250 rpm, in 2 ml LB Amp/Kan cultures inside 14 ml Falcon polystyrene tubes, pelleted, and resuspended in 100 µl fresh LB Amp/Kan. We spotted two 7.0 µl droplets of receiver cells on an LB Amp/Kan and 1 mM IPTG agar plate, and a 7.0-µl droplet of sender cells was spotted near the two receiver droplet. For this experiment, the plate was incubated at room temperature under the

Senders (cyan filter)

Receivers (yellow filter)

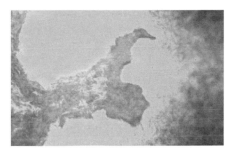

Overlay of images

Figure 7.24 Zoomed microscope fluorescence image of overlapping communicating cells (40× objective).

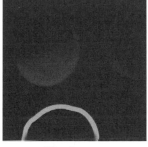

10 minutes

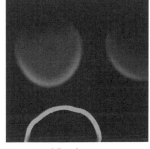

45 minutes

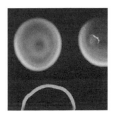

90 minutes

Figure 7.25 Time-series fluorescence images showing the response of receivers to communication from nearby senders on a plate.

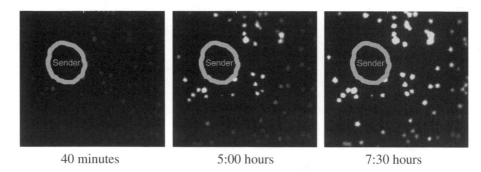

| 40 minutes | 5:00 hours | 7:30 hours |

Figure 7.26 Time-series fluorescence images illustrating the response of smaller colonies of receivers to communication from nearby senders on a plate.

microscope using a Plan Fluor $1 \times$ objective. The physical width of each image is approximately 7 mm.

First, a brightfield image was captured to record the location of the droplets. Then, a series of fluorescence images was captured at 1-min intervals using the green filter. The cyan semicircle in each fluorescence image is an artificial marker of the location of the sender droplet, superimposed based on the brightfield image.

The first fluorescence image was captured 10 min after spotting the droplets. It shows that the receiver cells closest to the senders have already started responding to the VAI message. The two subsequent images display an increase in fluorescence intensity due to the diffusion and accumulation of VAI. Based on the fluorescence response, VAI appears to diffuse at approximately 1 cm/h through the agar.

The third visual observation experiment captured a time-lapse series of images with smaller colonies of receivers and a smaller droplet of servers (Figure 7.26). Sender cells were grown overnight at 37°C, shaking at 250 rpm, in 2 ml LB Amp/Kan inside 14 ml Falcon polystyrene tubes. Receiver cells were picked from −80°C cell stock into 1000 μl LB Amp/Kan medium, then 20 μl was plated on LB Amp/Kan and incubated at 37°C for 7 h. Then 0.3 μl of sender cell were spotted, and brightfield and fluorescence images were captured as above. The three images show the communication gradient over time across the different receiver colonies.

Similar Signaling Systems

This section describes work that successfully isolates an important intercellular communication mechanism from a naturally occurring bacterial system, analyzes its components, and engineers its interfaces with standard genetic control and reporter mechanisms. While the work reported in this section captures

Table 7.1 Signaling systems similar to VAI: *N*-acyl-L-homoserine lactone autoinducers in bacteria.

Species	Relation to host	Regulates	I gene	R gene
Vibrio fischeri	Marine symbiont	Bioluminescence	*luxI*	*luxR*
Vibrio harveyi	Marine symbiont	Bioluminescence	*luxI*, M	*luxN,P,Q*
Pseudomonas aeruginosa	Human pathogen	Virulence factors	*lasI*	*lasR*
		Rhamnolipids	*rhlI*	*rhlR*
Yersinia enterocolitica	Human pathogen	?	*yenI*	*yenR*
Chromobacterium violaceum	Human pathogen	Violaceum production, hemolysin, exoprotease	*cviI*	*cviR*
Enterobacter agglomerans	Human pathogen	?	*eagI*	?
Agrobacterium tumefaciens	Plant pathogen	Ti plasmid conjugation	*traI*	*traR*
Erwinia caratovora	Plant pathogen	Virulence factors, carbapenem	*expI*	*expR*
Erwinia stewartii	Plant pathogen	Extracellular capsule	*esaI*	*esaR*
Rhizobium leguminosarum	Plant symbiont	Rhizome interactions	*rhiI*	*rhiR*
Pseudomonas aureofaciens	Plant beneficial	Phenazine production	*phzI*	*phzR*

one such communication mechanism, realistic genetically controlled developmental systems will require perhaps dozens of such signals. The *LasI/LasR* system from *Pseudomonas aeruginosa* [1], for example, appears to encode a similar regulatory system, but one that uses a different, and non–cross-reacting autoinducer and a different structure homologous to the Lux box. Table 7.1 lists additional signaling systems similar to VAI that could serve as potential communication signals [14]. Isolation and characterization of such additional communication mechanisms will allow the construction of more complex multicellular systems.

CONCLUSIONS

An important element in biocircuit design is genetic process engineering, a methodology for mutating the DNA encoding of existing genetic elements to achieve the desired input/output behavior for constructing reliable circuits of significant complexity. The optimized components we synthesized with this process exhibit the desired signal restoration and noise margins for reliable digital computation. We demonstrated the feasibility of digital computation in cells by building several operational *in vivo* digital logic circuits, each composed of three gates that have been optimized by genetic process engineering.

Because of the limits to single-cell circuit complexity, and to demonstrate coordinated behavior in cell aggregates, we also engineered intercellular communications. The mechanism relies on programmed enzymatic activity and chemical diffusions to carry messages, using DNA from the *Vibrio fischeri* lux operon. We built and characterized several circuits that integrate intracellular logic with lux operon-based intercellular communications.

References

[1] T. deKievit, P. C. Seed, J. Nezezon, L. Passador, and B. H. Iglewski. Rsal, a novel repressor of virulence gene expression in *Pseudomonas aeruginosa. J. Bacteriol.*, 181:2175–2184, 1999.

[2] A. Eberhard, A. L. Burlingame, C. Eberhard, G. L. Kenyon, K. H. Nealson, and N. J. Oppenheimer. Structural identification of autoinducer of *Photobacterium fischeri* luciferase. *Biochemistry*, 20(9):2444–2449, 1981.

[3] J. Engebrecht, K. H. Nealson, and M. Silverman. Bacterial bioluminescence: isolation and genetic analysis of the functions from *Vibrio fischeri. Cell*, 32:773–781, 1983.

[4] W. C. Fuqua and A. Eberhard. Signal generation in autoinduction systems: synthesis of acylated homoserine lactones by LuxI-type proteins. In G. M. Dunny and S. C. Winans, editors, *Cell-Cell Signaling in Bacteria*, pages 211–230. ASM Press, Washington, DC, 1999.

[5] W. C. Fuqua, S. Winans, and E. P. Greenberg. Quorum sensing in bacteria: the luxr-luxi family of cell density-responsive transcriptional regulators. *J. Bacteriol.*, 176:269–275, 1994.

[6] T. Gardner, R. Cantor, and J. Collins. Construction of a genetic toggle switch in *Escherichia coli. Nature*, 403:339–342, 2000.

[7] G. Green, S. R. Kain, and R. Angres. Dual color detection of cyan and yellow derivatives of green fluorescent protein using conventional microscopy and 35-mm photography. *Methods Enzymol.*, 327:89–94, 2000.

[8] T. F. Knight Jr. and G. J. Sussman. Cellular gate technology. In *Proceedings of UMC98: First International Conference on Unconventional Models of Computation*, pages 257–272, Auckland, NZ, January 1998. Springer-Verlag, Singapore.

[9] B. Muller-Hill. *The* lac *Operon: A Short History of a Genetic Paradigm*. Walter de Gruyter, 1996.

[10] M. Ptashne. *A Genetic Switch: Phage lambda and Higher Organisms*, 2nd ed. Cell Press and Blackwell Scientific Publications, Cambridge, 1986.

[11] E. G. Ruby and K. H. Nealson. Symbiotic association of *Photobacterium fischeri* with the marine luminous fish *Monocentris japonica:* a model of symbiosis based on bacterial studies. *Biol. Bull.*, 151:574–586, 1976.

[12] H. M. Shapiro. *Practical Flow Cytometry*, 3rd ed. Wiley-Liss, New York, 1995.

[13] A. M. Stevens and E. P. Greenberg. Transcriptional activation by LuxR. In G. M. Dunny and S. C. Winans, editors, *Cell-Cell Signaling in Bacteria*, pages 231–242. ASM Press, Washington, DC.

[14] S. Swift, P. Williams, and G. S.A.B. Stewart. *N*-acylhomoserine lactones and quorum sensing in proteobacteria. In G. M. Dunny and S. C. Winans, editors, *Cell-Cell Signaling in Bacteria*, pages 291–313. ASM Press, Washington, DC, 1999.

[15] R. Weiss. Cellular computation and communications using engineered genetic regulatory networks. PhD dissertation, Massachusetts Institute of Technology, 2001.

[16] R. Weiss, G. Homsy, and T. F. Knight Jr. Toward in-vivo digital circuits. In L. Landweber and E. Winfree, editors, *Evolution as Computation*, pages 275–295. Springer, Berlin, 2003.

8

The Biology of Integration of Cells into Microscale and Nanoscale Systems

Michael L. Simpson, Timothy E. McKnight,
Michael A. Guillorn, Vladimir I. Merkulov,
Gary S. Sayler, and Anatoli Melechko

In chapter 5 we focused on the informational interface between cells and synthetic components of systems. This interface is concerned with facilitating and manipulating information transport and processing between and within the synthetic and whole-cell components of these hybrid systems. However, there is also a structural interface between these components that is concerned with the physical placement, entrapment, and maintenance of the cells in a manner that enables the informational interface to operate. In this chapter we focus on this structural interface.

INTRODUCTION

Successful integration of whole-cell matrices into microscale and nanoscale elements requires a unique environment that fosters continued cell viability while promoting, or at least not blocking, the information transport and communication pathways described in earlier chapters. A century of cell culture has provided a wealth of insight and specific protocols to maintain the viability and (typically) proliferation of virtually every type of organism that can be propagated. More recently, the demands for more efficient bioreactors, more compatible biomedical implants, and the promise of engineered tissues has driven advances in surface-modification sciences, cellular immobilization, and scaffolding that provide structure and control over cell growth, in addition to their basic metabolic requirements. In turn, hybrid biological and electronic systems have emerged, capable of transducing the often highly sensitive and specific responses of cellular matrices for biosensing in environmental, medical, and industrial applications [20]. The demands of these systems have driven advances in cellular immobilization and encapsulation techniques, enabling improved interaction of the biological matrix with its environment while providing nutrient and respiratory requirements for prolonged viability of the living matrices [51].

Predominantly, such devices feature a single interface between the bulk biomatrix and transducer. However, advances in lithography, micromachining, and micro-/nanoscale synthesis provide broader opportunities for interfacing whole-cell matrices with synthetic elements. Advances in engineered, patterned, or directed cell growth are now providing spatial and temporal control over cellular integration within microscale and nanoscale systems [34]. Perhaps the best defined integration of cellular matrices with electronically active substrates has been accomplished with neuronal patterning. Topographical and physicochemical patterning of surfaces promotes the attachment and directed growth of neurites over electrically active substrates that are used to both stimulate and observe excitable cellular activity [64]. With the proper use of emerging techniques in the directed synthesis and assembly of nanoscale elements, the direct interface to smaller regions of individual cells and even subcellular molecular components are possible.

To date, the largest body of work is with microscale systems, as biosensors have incorporated cells with integrated circuits (IC) and microelectromechanical systems (MEMS). More recent efforts have pushed toward the molecular-scale interface between cells and synthetic components required for the types of systems ultimately envisioned here. In this chapter we review these efforts to integrate cells as components into microscale and nanoscale systems.

MICROSCALE SYSTEMS

Numerous demonstrations of the integration of cells into ICs, MEMS, or other microscale devices have been reported. These include microelectrode arrays for measuring action potentials from neuronal networks [58], a cell cartridge device that interfaces neurons or cardiomyocytes to an IC transducer [17], an IC transistor array that records electrical activity from on-chip cardiomyocyte cultures [33], an IC that both stimulates and records electrical activity from on-chip neurons [64], and many others. All these devices face similar issues regarding the incorporation of cells, which are very different from the other materials found in microscale devices. Our work has included extensive development of a whole-cell/IC sensor device we have coined the *bioluminescent bioreporter integrated circuit* (BBIC) [56, 57]. We present the BBIC as one example that illustrates the issues involved in the incorporation of the cells into microscale systems.

Bioluminescent Bioreporter Integrated Circuit

Many types of *lux* transcriptional gene fusions have been used to develop light-emitting bioreporter bacterial strains to sense the presence, bioavailability, and biodegradation of pollutants including toluene [3] and isopropylbenzene [55]. Analogous genetic approaches have also been reported for inducible heavy-metal detoxification and heavy-metal-resistance systems (including mercury [54]), and for response to heat shock [61] and oxidative stress [7]. In addition, genetically engineered Gram-positive bioreporters have been used to examine the efficacy of antimicrobial agents (decreased light equates to greater efficacy) [2, 15]. Eukaryotic bioreporters have also been generated to detect toxic compounds [1, 27], oxygen [23], ultraviolet light [9], and estrogenic and anti-estrogenic compounds [52]. Microorganisms with these *lux* gene fusions have been used extensively in biosensor devices [16, 20] by combining the cells with an appropriate light transducer.

Implicit in the use of a bioreporter strain for a biosensor is the assumption that the bioluminescent signal generated is directly related to the concentration of the target substance, most desirably in a selective and quantitative manner. In general, the *lux* reporter genes are placed under the regulatory control of inducible operons maintained in native plasmids, in plasmids with a broad range of hosts, or chromosomally integrated into the host strain. In these genetic systems, the target analyte or its degradation products act as the inducer of the bioluminescence genes and are responsible for selectivity and the resulting response. For example, *Pseudomonas fluorescens* HK44 is a bioreporter that produces light in the presence of naphthalene. This strain has two genetic operons positively regulated by NahR, a LysR-type protein (Figure 8.1).

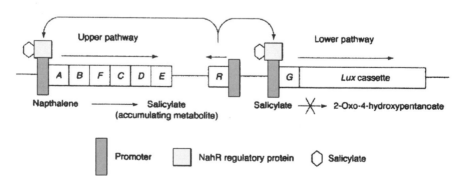

Figure 8.1 The two genetic operons positively regulated by NahR in the bioreporter *Pseudomonas fluorenscens* HK44 that produced light in the presence of naphthalene. The operon on the right contains the *lux* bioluminescence genes, while the other operon contains the genes responsible for the degradation of naphthalene to salicylate.

One of the operons contains the *lux* bioluminescence genes and the other genes responsible for the degradation of naphthalene to salicylate, the metabolic intermediate of naphthalene degradation. Both operons are induced when salicylate interacts with the regulatory protein NahR. Therefore, exposure of HK44 to either naphthalene or salicylate results in increased gene expression and increased bioluminescence. Initial studies in continuous cultures of *P. fluorescens* HK44 have demonstrated that the magnitude of the bioluminescence response correlated with the aqueous-phase concentration of naphthalene under pulsed-perturbation conditions [35]. Using growing cell cultures, a linear relationship to bioluminescence was found for both naphthalene and salicylate (correlation coefficient, r^2, of .990 for naphthalene and 0.991 for salicylate) over a concentration range of two orders of magnitude. Reproducible bioluminescence was observed not only in aqueous naphthalene samples but also in soil-slurry samples spiked with naphthalene, complex soil leachates and the water-soluble, components of jet fuel [30].

Combining these bioreporters with an integrated microluminometer forms a BBIC (Figure 8.2). We have implemented our microluminometer in a complementary metal oxide semiconductor (CMOS) technology.

This CMOS IC, shown in Figure 8.3, is described in detail elsewhere [57] and will not be presented further here because the details of the circuit design are not central to the incorporation of cells into microscale devices. However, one aspect of the BBIC that is of particular interest for the realization of highly integrated whole-cell devices is the packaging of the cells. We present this in some detail below.

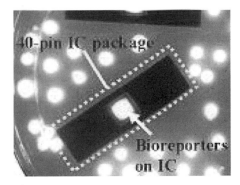

Figure 8.2 Bioluminescent bioreporter integrated circuits (BBICs). The integrated circuit (IC) is packaged in a 40-pin chip carrier and the bioreporters are positioned directly above the IC. All of the light for this photograph is provided by the bioluminescence (40-minute exposure).

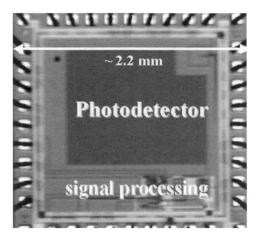

Figure 8.3 The integrated circuit portion of the BBIC. This integrated circuit was fabricated in a 0.5-μm, bulk, complementary, metal-oxide-semiconductor (CMOS) process. The chip measures approximately 2.2 mm on each side.

Bioreporter Entrapment

Entrapment of the bioreporters to the light-sensing portion of the integrated circuits presents unique challenges. The entrapment matrix must have mechanical strength, long shelf-life, resist dehydration, be nontoxic to the entrained cells and the environment, be nonpolluting, and have maximum cell retention. Further, the entrapment matrix must be porous to facilitate mass transfer of analytes, oxygen, and other essential nutrients. Because of the optical mode of communication common to many whole-cell devices, the matrix must be transparent or at least translucent and adhere to the light-sensing portion of the IC. The following subsection highlights the more widely used matrices.

Natural Polysaccharides

Alginates, agarose, and κ-carrageenan are natural polysaccharides. Alginates, produced by brown algae and certain bacterial species [11, 21, 22], are linear polymers of β(1,4)-D-mannuronic acid and α(1,4)-L-guluronic acid monomers. Alginates, depending on the algal source, vary in their composition, chain length, and arrangement and thus have variable properties. A cross-linking network is formed by the process of ionic gelation, bonding of Ca^{+2} ions with polyguluronic moieties of the polymer strands [11]. Sr^{+2} can be substituted for the Ca^{+2}, improving the mechanical stability of the matrix [31]. Alginates with a higher guluronic acid content bind more Ca^{+2} resulting in a stronger matrix. Ionic bonding occurs immediately and is usually complete in 30–60 min depending on the cell, alginate, and Ca^{+2} concentrations [24]. Ionic gelation of the alginates is a temperature-independent process; working temperatures range from 0°C to 80°C. For encapsulation, cells are mixed with alginate (1–8%) and are usually extruded through a syringe into a Ca^{+2} (or Sr^{+2}) solution to generate beads. A 21-gauge syringe produces alginate beads of approximately 3-mm diameter [37]. Additional structural integrity can be achieved by coating the alginate beads with poly-L-lysine [51].

The carrageenans are produced by red algae. There are three types (ι, λ, κ), all of which have a common backbone of alternating β(1,3)-D-galactose and α(1,4)-D-galactose [11]. The difference in the three types is the degree and location of sulfonation on each sugar moiety. In contrast to the alginates, gelation is temperature dependent, but the strength of the matrix can be adjusted with the addition of K^+ or Al^{+3} ions [26]. κ-Carrageenan (2–5% in the presence of 0.1–0.3 M KCl) is the most widely used because of its firmer gelling properties [36]. For immobilization, κ-carrageenan is heated to 80°C to dissolve the polysaccharide and then cooled to 45°C. The cell suspension is also heated to 45°C and mixed with the κ-carrageenan. Similar to the alginate, the cell κ-carrageenan mixture is extruded through a syringe into cold KCl solution to make small beads [11]. The obvious disadvantage of this method is the heating and rapid cooling of the cells. Any cell culture would have to tolerate 45°C for a short period of time. There are numerous examples in the literature of cell immobilization in these natural polysaccharides [12, 47–49].

Sol gel

Sol-gel matrices are porous wet gels obtained by the hydrolysis and condensation/polymerization of metal and semimetal alkoxides (SiO_2 materials) [4, 29]. Sol-gel characteristics include thermal stability, controllable surface area and pore size, transparency, and nontoxicity [4]. Microbial cells as well as enzymes have been immobilized by sol-gels [18, 19, 32, 63]. Armon et al. [4] achieved film thicknesses of 0.1–0.2 mm. Chia et al. [14] reported that sol-gel films of 0.1 μm can be achieved. However, our experience has shown that sol-gel thin-films (100 μm) are extremely fragile.

Acrylate-Vinyl Acetate Copolymers

A more recent approach is the use of latex (acrylate-vinyl acetate copolymers) for the immobilization of bacterial cells [39, 40, 41]. The advantages of these copolymers include mechanical strength at very thin film thicknesses (10–100 μm), high cell retention, and higher mass cell loadings compared to natural polysaccharides. Lyngberg and co-workers [39] demonstrated latex films with microbial cell densities as high as 50 vol% and 2×10^{11} cells/cm^3 of coating volume. These thin layers minimize mass transfer limitations of analytes (target compound, nutrients, O_2) to the bioreporter. Additionally, photons produced by the bioreporter have a shorter path and fewer obstructions to the light-sensing portion of the BBIC.

Lyngberg et al. [41] immobilized *E. coli* HB101 (*mer-lux*) as part of a mercury biosensor. A preparation of 1.2 g cell paste was resuspended in 0.3 ml 50% glycerol and 1 ml Rovace SF091 (Rohm and Haas, Philadelphia, PA). This was spread onto a polyester support using a 26-mil-wire Mayer bar to a thickness of 30 μm. After curing this first layer, a second layer (47 μm) of latex was applied as a top coat. Using fresh cells in this biosensor, 0.1–10,000 nM $HgCl_2$ was detected. These latex films were also stable at −20°C for 3 months.

A disadvantage of the latex approach is the requirement for rigid curing protocols. To maintain porosity but promote cell retention, the latex spheres must be partially cured under controlled humidity and temperature. Full curing will yield a solid film with no porosity and deprives the bioreporter cells of water.

Cellulose-binding Domain

Cellulose-degrading bacteria produce an extracellular enzyme complex called cellulases [6, 8]. One feature of the cellulases is the cellulose-binding domain (CBD), a subunit that physically binds to cellulose. In some species, binding by the CBD is irreversible [60]. CBD has been used as an affinity tag for the purification/immobilization of proteins and antibodies [5, 50, 53, 59] as well as for immobilization of whole cells [62]. CBD incorporation into bacterial cells and using cellulose as a support matrix has several advantages over the previous methods: (1) cellulose is an abundant, inexpensive material that comes in many forms; (2) a uniform monolayer of cells can be achieved putting the bioluminescent bioreporters closer to the light-sensing portion of the integrated circuit; (3) binding is rapid and essentially irreversible; (4) there are no prolonged curing steps or curing agents that may inhibit cellular activity; and (5) CBDs are stable and resistant to denaturation and proteolytic degradation [60].

Current Approaches

Previously our work has focused on either alginate or agarose for immobilizing bioreporters [31]. The advantage of alginate or agarose beads is that they are easy to prepare and large quantities can be made and stored at 4°C for long periods (weeks to months) with no loss of activity. However, these natural polysaccharides may not be suitable for IC applications. Most preparations are by extrusion of cells in alginate from a syringe into a solution or onto a solid surface. This does not generate films of uniform thickness, and the films are relatively thick (hundreds of microns).

We have begun to explore the use of latex as an immobilization agent similar to Lyngberg et al. [39, 41]. As previously mentioned, the latex can be rolled into ultrathin sheets (10–100 μm) of uniform thickness. Using *Pseudomonas*

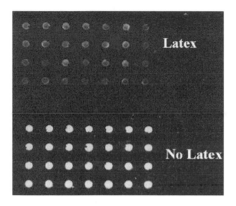

Figure 8.4 Comparison of salicylate-induced *P. fluorescens* 5RL cells with and without latex encapsulation. Bioluminescence from encapsulated cells appears somewhat attenuated.

fluorescens 5RL, a naphthalene-*lux* bioreporter, bioluminescence production was compared between immobilized and free cells. Rovace SF-091 (Rohm and Haas, Philadelphia, PA) was used to determine toxicity effects on *Pseudomonas fluorescens* 5RL. Uninduced, log-phase cells ($OD_{546} = 0.35$) were washed with mineral salts buffer (MSM) and resuspended in MSM + 15% glycerol. Cells were spotted onto duplicate cellulose membranes and excess moisture was allowed to dry. The cell spots were coated with the latex mixture at full strength and dried overnight at 5°C and at 40% humidity.

We determined viability of cells by laying the membranes on LB broth supplemented with sodium salicylate (100 ppm) and qualitatively monitoring light production. Cells coated with Rovace SF-091 had diminished bioluminescence compared to the control membrane (Figure 8.4).

A growing-cell assay experiment was performed in a microtiter plate using *P. fluorescens* 5RL in the presence and absence of Rovace SF-091 to further quantify the reduction in bioluminescence. The purpose of the experiment was to demonstrate the inhibitory effect of the latex on the bioluminescent response of the cells. The cells were immobilized on a nylon membrane (BIOTRANS Nylon Membranes, ICN Biomedical, Aurora, OH) with a pore size of 0.45 μm using a dot-blot filter apparatus (Bio-Rad, Hercules, CA). Cells were placed in individual wells in the dot-blot apparatus and transferred to the membrane by applying a vacuum. The number of cells per patch was calculated to be 1×10^7 CFU. The vacuum was maintained until the cell patches appeared dry. One-half of the membrane was placed on a slanted surface, and the latex was poured over the membrane until it was covered. Excess latex driped off the membrane leaving a thin latex film over the membrane. The latex was dried at 5°C and at 40% humidity overnight. The other half of the membrane was stored overnight at the same temperature and humidity. The next day, the individual patches were cut and placed on top of a sterile sponge (3M ScotchBrite) slightly larger than the patches. The sponge pieces and patches were placed in the wells

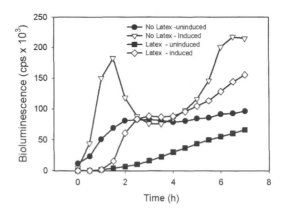

Figure 8.5 Salicylate induction of unencapsulated and Rovace SF-091 encapsulated *P. fluorescens* 5RL cells.

of a 24-well microtiter plate. Luria-Bertani medium supplemented with sodium salicylate at a concentration of 100 ppm and tetracycline was added to each well. The plate was placed in a Wallac Luminometer, and bioluminescence readings were recorded every 30 min for 7 h.

In the presence of salicylate and no latex, *P. fluorescens* 5RL exhibited its classical response with a sharp increase in light production in the first 90 min, followed by a gradual reduction over 90–240 min and then another increase in light production (Figure 8.5). In the presence of latex, bioluminescence was not observed until 90 min, followed by a plateau at 180–240 min. After 240 min, there was a steady increase in bioluminescence. The cause of the lag phase is not clear. It may be due to diffusion of the inducer through the latex matrix or an unidentified physiological effect on the cells themselves.

Although the physiological effect of latex on bioluminescence bioreporter responses requires more study, the use of latex as an immobilization agent has many advantages for the encapsulation of cells deployed in microscale or nanoscale systems. Because of the burden of maintaining cell viability, cell encapsulation will remain an important topic in the device science of cells.

NANOSCALE SYSTEMS

Interfacing with cells on the microscale is sufficient for interaction with single genetic regulatory systems through reporter genes or the stimulation and recording of action potentials. However, broadly interfacing with genetic circuits and networks with the goal of fully using the functional complexity of these systems in engineered devices requires an interface on the same size scale as the biomolecular machines that carry out these functions. Synthetic nanostructures are at the confluence of the smallest of human-made devices and the largest of these biomolecules, and the controlled synthesis followed by the directed assembly of nanoscale synthetic components can lead to nanostructured substrates that interact with cells at the molecular level.

One example of a hybrid nanostructured substrate/whole-cell device comes from the field of biomedical devices in the area of transplanted cells for the treatment of hormone deficiencies arising from diseases such as Type I diabetes. The transplant of islet cells from other animals into human diabetics (xeno-transplantation) is seen as a promising approach to the careful control of blood sugar levels; these cells can both sense glucose and respond by producing insulin in a closed-loop system. For this strategy to work, the transplanted cells must be isolated from the patient's immune response to maintain their viability. Recent work has focused on the development of biocompatible nanoporous capsules that allow the free flow of glucose, insulin, and other essential nutrients for the islets, yet inhibit the passage of larger entities associated with an immune response [38]. The result is a hybrid device where all the information processing and actuation are performed by the cells, and a synthetic nanostructured substrate manipulates the molecular communication with these cells to allow their functionality while maintaining their viability.

In the example above, the genetic circuits of the cells were evolved to perform the required functions, and the nanostructured portion of the synthetic device was only required to manipulate molecular communication through size discrimination. To use the complex functionality of the cells as envisioned here, a more complete molecular connection between the cells and substrate is required. Toward this goal, luminescent semiconductor quantum dots have been derivatized with biomolecules and used as fluorescent probes in intracellular assays [10, 13]. Although this enhances communication from the molecular processes of cells, it does not allow communication to or control of these processes. However, electronic control over the local hybridization behavior of DNA molecules by inductive coupling of a radio-frequency magnetic field to metal nanocrystals covalently linked to DNA was recently reported [28]. To move further along this path of controlling molecular processes and bidirectional communication between cellular components and nanostructured substrates requires the controlled synthesis of nanoscale elements that interface with cells and their integration into microscale devices.

We have recently begun investigating the controlled synthesis of individually addressable, vertically aligned carbon nanofiber (VACNF) elements (Figure 8.6), focused on the synthesis, surface modification, and chemical derivitization of these nanoscale structures. VACNFs are synthesized in a plasma-enhanced chemical vapor deposition process and grow perpendicular to the substrate from metal catalyst particles. We have demonstrated that VACNF synthesis is an extremely flexible method for the directed assembly of nanoscale features within

Figure 8.6 Vertically aligned carbon nanofibers (VACNFs). (Left) A scanning electron micrograph of an individual isolated VACNF. (Center) A random forest of VACNFs grown from an unpatterned catalyst. (Right) A deterministic array of VACNFs grown from lithographically defined catalyst dots.

microscale and larger structures. VACNF location, shape, length, tip diameter, and, to some extent, chemical composition can be controlled by selection of substrate or plasma growth conditions [43–46]. Standard lithographic and microfabrication techniques have been used to incorporate VACNFs into microscale structures to provide individual addressability, electrochemical passivation, and the activation of discrete nanoscale electrochemically active areas at the tips or along defined lengths of each independent element (Figure 8.7) [25].

Of importance for the types of hybrid systems envisioned here, VACNFs provide the potential for the realization of molecular-scale informational and structural interfaces to cells or cellular communities. As an example, we have recently demonstrated that periodic arrays of VACNFs may be implemented for biochemical manipulation within the intracellular, and even nuclear, domains of mammalian cellular matrices [42]. By modifying the surface of nanofibers with

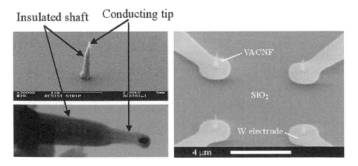

Figure 8.7 Standard microfabrication techniques were used to incorporate vertically aligned carbon nanofibers (VACNFs) into microscale structures to provide individual addressability, electrochemical passivation, and the activation of discrete nanoscale electrochemically active areas at the tips.

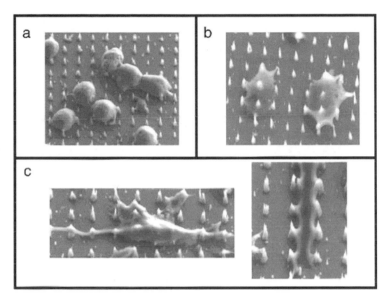

Figure 8.8 Scanning electron micrographs of CHO cells after (a) centrifugation of cells onto vertically aligned carbon nanofiber array (b) press of vertically aligned carbon nanofibers into the cells, and (c) culture for 48 h after steps a and b.

either adsorbed or covalently attached plasmid DNA, fiber arrays can be pressed into (i.e., interfaced with) viable cells (Figure 8.8) on a massively parallel basis to deliver genetic material or other macromolecules into the cells.

Once interfaced to nanofibered substrates, cells were observed to recover and proliferate on the fibered substrate and to express the delivered gene for extended time periods (Figure 8.9). In addition to demonstrating a promising method for *massively parallel* microinjection of cellular/tissue matrices, the results of these experiments indicated that plasmid DNA can be expressed within CHO cells even while covalently tethered to a penetrant nanofiber scaffold.

As these plasmid DNAs are immobilized on the nanofiber, their potential for genomic insertion is likely reduced, and data indicated that segregation of these bound plasmids to progeny is rare. Thus, the cells interfaced with the nanofibers can be provided with expressable, exogenous DNA that is non-inheritable (Figure 8.10). Furthermore, these plasmids can be physically removed from the cell along with the nanofiber scaffold, providing a unique strategy for temporal control of gene expression, the basis of all cellular information processing.

In addition to gene delivery and physical control of gene expression, nanofibers may be implemented as electrochemical probes that can be used for cell-to-chip and chip-to-cell communication, either by direct electrochemical methods (oxidation/reduction processes) or by electrochemical/electrical/thermal manipulation of nanofiber-bound molecules, including DNA and

Figure 8.9 Fluorescent micrograph of a large CHO cell colony 22 days after cellular integration with a nanofiber array loaded with a plasmid containing a green fluorescent protein (GFP) gene. The fluorescence seen in this image indicates that the CHO cells are viable and still expressing the GFP gene more than 3 weeks after the initial transfection with vertically aligned carbon nanofibers.

enzymes. The device in Figure 8.7 is a nanoscale version of the carbon microelectrode described earlier, and we have demonstrated its functionality as a probe where only the extreme nanoscale tip of the fiber is electrochemically active (Figure 8.11) [25]. In combination with the ability to introduce chemically functionalized nanofibers into cells on a massively parallel basis, such nanoscale electrodes may provide unique approaches for electronically monitoring and controlling molecular-scale phenomena within cells.

VACNF arrays may also provide unique approaches to cellular immobilization and scaffolding. Because they extend out of the plane of the substrate,

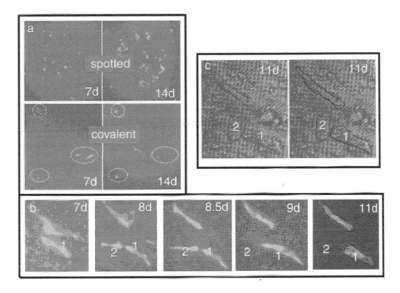

Figure 8.10 Fluorescent micrographs of CHO cells expressing a vertically aligned carbon nanofiber (VACNF)-delivered green fluorescent protein (GFP) gene. (a) Time-lapse images of GFP-expressing cells from (top) spotted (nonspecifically adsorbed plasmid) nanofibers, and (bottom) nanofibers with covalently linked plasmid. The spotted samples tend to produce colonies of cells from initial transfectants, whereas the covalently linked samples tend to maintain a constant number of GFP-expressing cells. (b) A time-lapse sequence of a typical cell division on a covalently linked plasmid VACNF array indicating that the plasmid DNA is not segregated to progeny cells. Within 1 day after cell division, no fluorescence is observed from the daughter cell not retained on the nanofiber. (c) Brightfield image demonstrates that the daughter cell still resides adjacent to the mother cell. This division and subsequent loss of GFP expression in the daughter cells for covalently linked plasmid VACNF samples leads to a constant number of GFP-expressing cells over long periods of time, as indicated in the lower two panels of part a.

VACNFs provide the means to elevate cellular matrices above the substrate, enhancing diffusion of molecular species while still providing electrical and electrochemical interconnectivity with the suspended matrix (Figure 8.12).

CONCLUSIONS

The successful incorporation of whole cells as functional elements in microscale and nanoscale devices requires a unique environment that fosters continued cell viability and provides for information transport between the synthetic and bio-logical substrates. There has been significant progress in this area from many investigators, including the integrated circuit-based approach and the numerous

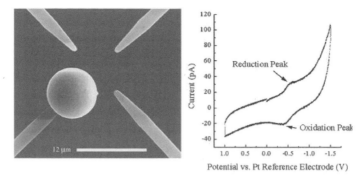

Figure 8.11 Demonstration of the electrochemical activity of the nanoscale probes of Figure 8.7. (Left) Scanning electron micrograph of a vertically aligned carbon nanofiber (VACNF) array following electrode position of gold on the lower-left probe. (Right) Cyclic voltammetry data for a VACNF-based electrochemical probe device. These curves were taken in 1 mM ruthenium hexamine trichloride buffered by a 0.1 M KCl solution using platinum reference and counter electrodes. The curves show clear oxidation and reduction peaks; the low level of measured faradic current indicates that only a small area of the device is electrochemically active, presumably the VACNF probe tip.

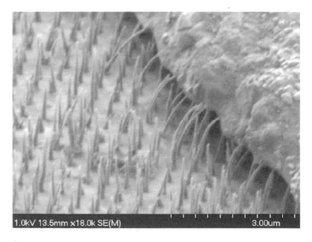

Figure 8.12 A biofilm of *Pseudomonas* cells grown on a two-dimensional array of carbon nanofibers. Vertically aligned carbon nanofiber arrays provide unique approaches to cellular scaffolding that combine mechanical strength of nanofiber networks with the potential for individually addressing each scaffold element.

cell encapsulation methods described here. Although these whole-cell biosensors provide one of the practical successes of molecular-scale devices, these devices use only a small portion of the full functionality of the cells. Accessing a greater portion of this functionality requires a more comprehensive coupling to the informational machinery of the cells. We have shown that arrays of carbon nanofiber elements may be functionally integrated within cells that remain viable after VACNF insertion. This successful integration provides an intracellular biochemical interface that is a critical enabling step both as a potential communication pathway between the synthetic and biological substrates in hybrid devices and in the realization of hardware tools that couple modeling and experiment in genetic circuit and network exploration and design. At this time, the exploration of such interfaces between engineered nanostructures and biomolecular processes of cells is in its infancy. However, since this coupling of nanotechnology and biotechnology holds such great promise, there is likely to be significant near-term progress in this area. For example, the functioning of the hybrid combination of cell and VACNFs was demonstrated by long-term gene expression from nanofiber-bound plasmid molecules, indicating that other functional properties of VACNFs can be used during intracellular deployment without affecting cell viability. Perhaps previously demonstrated VACNF electrochemical probes [25] could be deployed within cells, or electronically or heat-mediated methods for binding or releasing the active coding regions of the delivered plasmid to the nanofiber could provide specific temporal control of gene expression. In any event, as nanoscale science, engineering, and technology continue to push synthetic devices down to the molecular scale, our ability to interject these devices into the flow of information within cells will grow. And, as our ability to connect to these cellular information pathways grows, so to does our ability to harness this functionality for engineered devices.

References

[1] J. W. Anderson et al. A biomarker, p450 rgs, for assessing the potential toxicity of organic compounds in environmental samples. *Environ. Toxicol. Chem.*, 14:1159–1169, 1995.

[2] P. W. Andrew and I. S. Roberts. Construction of a bioluminescent mycobacterium and its use for assay of antimycobacterial agents. *J. Clin. Microbiol.*, 31:2251–2254, 1993.

[3] B. Applegate et al. *Pseudomonas putida* b2: a tod-lux bioluminescent reporter for toluene and trichloroethylene co-metabolism. *J. Ind. Microbiol.*, 18:4–9, 1997.

[4] R. Armon et al. Sol-gel applications in enviromental biotechnology. *J. Biotechnol.*, 51:279–285, 1996.

[5] Z. Assouline, D. G. H. Shen, and R. A. J Kilburn. Production and properties of a factor x-cellulose-binding domain fusion protein. *Protein Eng.*, 6:787–792, 1993.

[6] P. Beguin and J. P. Aubert. The biological degradation of cellulose. *FEMS Microbiol. Rev.*, 13:25–58, 1994.

[7] S. Belkin et al. Oxidative stress detection with *Escherichia coli* harboring a katg'::
lux fusion. *Appl. Environ. Microbiol.*, 62:2252–2256, 1996.

[8] M. K. Bhat. Cellulases and related enzymes in biotechnology. *Biotechnol. Adv.*, 18:355–383, 2000.

[9] C. Biberger and E. V. Angerer. 2-Phenylindoles with sulfur containing side chains. estrogen receptor affinity, antiestrogenic potency, and antitumor activity. *J. Steroid Biochem. Mol. Biol.*, 58:31–43, 1996.

[10] M. Bruchez Jr. et al. Semiconductor nanocrystals as fluorescent biological labels. *Science*, 281:2013–2016, 1998.

[11] M. B. Cassidy, H. Lee, and J. T. Trevors. Environmental applications of immobilized microbial cells: a review. *J. Ind. Microbiol.*, 16:79–101, 1996.

[12] P. Caunt and H. A. Chase. Degradation of n-valeric acid by alginate-entrapped alcaligenes denitrificans. *Appl. Microbiol. Biotechnol.*, 25:453–458, 1987.

[13] W. C. W. Chan and S. M. Nie. Quantum dot bioconjugates for ultrasensitive non-isotopic detection. *Science*, 281:2016–2018, 1998.

[14] S. Chia et al. Patterned hexagonal arrays of living cells in sol-gel silica films. *J. Am. Chem. Soc.*, 122:6488–6489, 2000.

[15] R. C. Cooksey et al. A rapid method for screening antimicrobial agents for activities against a strain of *Mycobacterium tuberculosis* expressing firefly luciferase. *Antimicrob. Agents Chemother.*, 37:1348–1352, 1993.

[16] S. Daunert et al. Genetically engineered whole-cell sensing systems: coupling biological recognition with reporter genes. *Chem. Rev.*, 100(7):2705–2738, 2000.

[17] B. D. DeBusschere and G. T. A. Kovacs. Portable cell-based biosensor system using integrated cmos cell-cartridges. *Biosens. Bioelectr.*, 16:543–556, 2001.

[18] C. Dosoretz, R. Armon, J. Starosvetzky, and N. Rothschild. Entrapment of parthion hydrolase from *Pseudomonas* spp. in sol-gel glass. *J. Sol-Gel Sci. Technol.*, 7:7–11, 1996.

[19] L. M. Ellerby et al. Encapsulation of proteins in transparent porous silicate glasses prepared by sol-gel method. *Science*, 255:1113–1115, 1992.

[20] D'Souza S. F. Microbial biosensors. *Biosens. Bioelectron.*, 16(6):337–353, 2001.

[21] W. F. Fett, S. F. Osman, L. Frishman, and T. S. Siebles III. Alginate production by plant-pathogenic pseudomonads. *Appl. Environ. Microbiol.*, 52:466–473, 1986.

[22] W. F. Fett and C. Wijey. Yields of alginates produced by fluorescent pseudomonads in batch culture. *J. Ind. Micobiol.*, 14:412–415, 1995.

[23] D. Filatov, S. Bjorklund, E. Johansson, and L. Thelander. Induction of the mouse ribonucleotide reductase r1 and r2 genes in response to DNA damage by UV light. *J. Biol. Chem.*, 271:23698–23704, 1996.

[24] D. D. Gilson, A. Thomas, and F. R. Hawkes. Gelling mechanism of alginate beads with and without immobilized yeast. *Process Biochem.*, 25:104–108, 1990.

[25] M. A. Guillorn et al. Individually addressable vertically aligned carbon nanofiber-based electrochemical probes. *J. Appl. Phys.*, 91(6):3824–3828, March 2002.

[26] K. B. Guiseley. Chemical and physical properties of algal polysaccharides used for cell immobilization. *Enzyme Microbiol. Technol.*, 11:706–716, 1989.

[27] M. Gupta and E. Goldwasser. The role of the near upstream sequence in hypoxia-induced expression of the erythropoietin gene. *Nucleic Acids Res.*, 24:4768–4774, 1996.

[28] K. Hamad-Schifferli et al. Remote electronic control of DNA hybridization through inductive coupling to an attached metal nanocrystal antenna. *Nature*, 415:152–155, 2002.

[29] Y. Haruvy, A. Heller, and S. E. Webber. Sol-gel preparation of optically clear supported thin-film glasses embodying lase dyes. In T. Bein, editor, ACS Symposium Series No. 499, *Supramolecular Architecture: Synthetic Control in Thin Films and Solids*, pages 405–424. American Chemical Society, Washington, DC, 1992.

[30] A. Heitzer et al. Optical biosensor for environmental on-line monitoring of naphthalene and salicylate bioavailability with an immobilized bioluminescent catabolic reporter bacterium. *Appl. Environ. Microbiol.*, 60:1487–1494, 1994.

[31] A. K. Heitzer et al. Optical biosensor for environmental on-line monitoring of naphthalene and salicylate bioavailability with an immobilized bioluminescent catabolic reporter bacterium. *Appl. Environ. Microbiol.*, 60:1487–1494, 1994.

[32] L. Inama, S. Dire, G. Carturan, and A. Cavazza. Entrapment of viable microorganisms by sio2 sol gel layers on glass surfaces: trapping, catalytic performance and immobilization durability of saccharomyces cerevisiae. *J. Biotechnol.*, 30:197–210, 1993.

[33] S. Ingebrandt, C. Yeung, M. Krause, and A. Offenhausser. Cardiomyocyte-transistor-hybrids for sensor application. *Biosens. Bioelectr.*, 16:565–570, 2001.

[34] D. R. Jung et al. Topographical and physicochemical modification of material surface to enable patterning of living cells. *Crit. Rev. Biotechnol.*, 21(2):111–154, 2001.

[35] J. M. H. King et al. Novel bioluminescent reporter technology for naphthalene exposure and biodegradation. *Science*, 249:778–781, 1990.

[36] J. Klein and K. D. Vorlop. Immobilization techniques-cells. In M. Moo-Young, C. L. Cooney, and A. E. Humphrey, editors, *Comprehensive Biotechnology: Engineering Considerations*, pages 203–224. Pergamon Press, Oxford, 1985.

[37] E. D. Lancy and O. H. Tuovinen. Ferrous ion oxidation by thiobacillus ferrooxidans immobilized in calcium alginate. *Appl. Microbiol. Biotechnol.*, 20:94–99, 1984.

[38] L. Leoni and T. A. Desai. Nanoporous biocapsules for the encapsulation of insulinoma cells: biotransport and biocompatibility considerations. *IEEE Trans. Biomed. Eng.*, 11, 2001.

[39] O. K. Lyngberg et al. A patch coating method for preparing biocatalytic films of *Escherichia coli*. *Biotechnol. Bioeng.*, 62:44–55, 1999.

[40] O. K. Lyngberg et al. Engineering the microstructure and permeability of thin multiplayer biocatalytic coatings containing *E. coli*. *Biotechnol. Prog.*, 17:1169–1179, 2001.

[41] O. K. Lyngberg, D. J. Stemke, L. E. Scriven, and M. C. Flickinger. A simple single use luciferase based mercury biosensor using latex film immobilized *Escherichia coli* hb101. *J. Ind. Microbiol. Biotechnol.*, 23:668–676, 1999.

[42] T. E. McKnight et al. Intracellular integration of synthetic nanostructures with

viable cells for controlled biomechanical manipulation. *Nanotechnology*, 14:551–556, 2003.

[43] V. I. Merkulov et al. Patterned growth of individual and multiple vertically aligned carbon nanofibers. *Appl. Phys. Lett.*, 76(24):3555–3557, 2000.

[44] V. I. Merkulov et al. Shaping carbon nanostructures by controlling the synthesis process. *Appl. Phys. Lett.*, 79:1178–1180, 2001.

[45] V. I. Merkulov et al. Sharpening of carbon nanocone tips during plasma-enhanced chemical vapor growth. *Chem. Phys. Lett.*, 350:381–385, 2001.

[46] V. I. Merkulov et al. Effects of spatial separation on the growth of vertically aligned carbon nanofibers produced by plasma-enhanced chemical vapor deposition. *Appl. Phys. Lett.*, 80:476–478, 2002.

[47] P. Mussenden, T. Keshavarz, G. Saunders, and C. Burke. Physiological studies related to the immobilization of *Penicillium chrysogenum*. *Enzyme Microb. Technol.*, 15:2–7, 1993.

[48] R. Nandi, P. K. Bhattacharyya, A. N. Bhaduri, and S. Sengupta. Synthesis and lysis of formate by immobilized cells of *Escherichia coli*. *Biotechnol. Bioeng.*, 39:775–780, 1992.

[49] M. S. Nawaz, W. Franklin, and C. E. Cerniglia. Degradation of acrylamide by immobilized cells of a *Pseudomonas* sp. and *Xanthomonas maltophilia*. *Can. J. Microbiol.*, 39:207–212, 1992.

[50] E. Ong et al. Enzyme immobilization using a cellulose-binding domain: properties of a b-glucosidase fusion protein. *Bio/Technol.*, 7:604–607, 1989.

[51] J. K. Park and H. N. Chang. Microencapsulation of microbial cells. *Biotechnol. Adv.*, 18:303–319, 2000.

[52] M. Pons, D. Gagne, J. C. Nicolas, and M. Mehtali. A new cellular model of response to estrogens: a bioluminescent test to characterize (anti) estrogen molecules. *BioTechniques*, 9:450–459, 1990.

[53] R. D. Richins, A. Mulchandani, and W. Chen. Expression, immobilization, and enzymatic characterization of cellulose-binding domain-organo phosphorous hydrolase fusion enzymes. *Biotechnol. Bioeng.*, 69:591–596, 2000.

[54] O. Selifonova, R. S. Burlage, and T. Barkay. Bioluminescent sensors for detection of bioavailable hg(ii) in the environment. *Appl. Environ. Microbiol.*, 59:3083–3090, 1993.

[55] O. V. Selifonova and R. W. Eaton. Use of an ipb-lux fusion to study regulation of the isopropylbenzene catabolism operon of *Pseudomonas putida* re204 and to detect hydrophobic pollutants in the environment. *Appl. Environ. Microbiol.*, 62:778–783, 1996.

[56] M. L. Simpson et al. Bioluminescent-bioreporter integrated circuits form novel whole-cell biosensors. *Trends Biotech.*, 16:332–338, 1998.

[57] M. L. Simpson et al. An integrated cmos microluminometer for low-level luminescence sensing in the bioluminescent bioreporter integrated circuit. *Sens. Act. B*, 72(2):135–141, 2001.

[58] D. A. Stenger et al. Detection of physiologically active compounds using cell-based biosensors. *Trends Biotechnol.*, 19(8):304–309, 2001.

[59] P. Tomme et al. An internal cellulose-binding domain mediates adsorption on an engineered bifunctional xylanase/cellulase. *Protein Eng.*, 7:117–123, 1994.

[60] P. Tomme et al. Characterization and affinity applications of cellulose-binding domains. *J. Chromatography B.*, 715:283–296, 1998.

[61] T. K. Van Dyk et al. Rapid and sensitive pollutant detection by induction of heat shock gene-bioluminescence gene fusions. *Appl. Environ. Microbiol.*, 60:1414–1420, 1994.

[62] A. A. Wang, A. Mulchandani, and W. Chen. Whole-cell immobilization using cell surface-exposed cellulose-binding domain. *Biotechnol. Prog.*, 17:407–411, 2001.

[63] H. H. Weetall et al. Bacteriorhodopsin immobilized in sol-gel glass. *Biochim. Biophys. Acta*, 1142:211–213, 1993.

[64] G. Zeck and P. Fromherz. Noninvasive neuroelectronic interfacing with synaptically connected snail neurons immobilized on a semiconductor chip. *Proc. Natl. Acad. Sci. USA*, 98(18):10457–10462, 2001.

Part III

Computation in Ciliates

9

Encrypted Genes and Their Assembly in Ciliates

David M. Prescott and Grzegorz Rozenberg

Maintenance of normal cell function and structure requires some level of stability of the cell's DNA—at least the DNA that makes up the genes of the cell. In most eukaryotes most of the DNA in the genome does not encode genes and has no known function beyond forming long spacers between successive genes. For example, the gene density in the germline (micronuclear) genome of stichotrich ciliates (formerly referred to as hypotrich ciliates) is very low; only a few percent of the DNA encodes the approximately 27,000 different genes, and more than 95% is spacer DNA. Powerful DNA repair systems guard the stability both of nongene and gene DNA in contemporary cells, protecting it against mutagenesis. Although species survival depends on DNA stability, cell evolution requires changes in DNA. Presumably, there is a balance between instability of DNA that allows evolution and a stability that protects species from mutational extinction. Could cells evolve strategies that change the balance, allowing a greater rate of DNA change (gene evolution) without jeopardizing species survival? The stichotrichs may, in fact, have evolved such a mechanism, dramatically modifying their germline DNA during evolution to facilitate creation of new genes without reducing the level of cell survival. The modifications of germline DNA in ciliates, in turn, require dramatic DNA processing to convert germline DNA into somatic DNA during the life cycle of the organisms.

INTRODUCTION

The ciliate strategy rests on the evolution of nuclear dimorphism: the inclusion both of a germline nucleus (micronucleus) and a somatic nucleus (macronucleus) in the same cell (Figure 9.1; for a general review, see Prescott [6, 7]). Like the example in Figure 9.1, most stichotrich species contain two or more micronuclei and two or more macronuclei per cell.

The multiple micronuclei are genetically identical to each other, and the multiple macronuclei are genetically identical; these multiplicities of nuclei have no bearing on the issues addressed in this chapter. The micronucleus is used only in cell mating, and its genes are silent. Hence, micronuclear genes do not support the maintenance, growth, or division of the cell. The genes in the macronucleus are actively transcribed into the mRNA molecules required to maintain cell structure and function.

In spite of their vastly different functions, the micronucleus and macronucleus are closely related, a relationship that is evident in the sexual phase of the ciliate life cycle. Stichotrich ciliates make a living by ingesting other organisms, such as other ciliates, unicellular alga, and bacteria. When such food organisms are available, stichotrich ciliates proliferate rapidly, typically undergoing cell division every 6–8 h. When starved for food, ciliates may enter a sexual phase, in which they mate in pairs. Mating begins with the joining of two cells by a cytoplasmic channel (Figure 9.2).

Cell joining is quickly followed by meiosis of the micronuclei in each cell, in which four haploid micronuclei are formed from each diploid micronucleus

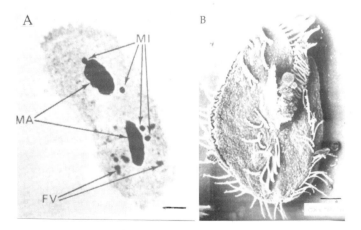

Figure 9.1 (a) A light microscope photograph of a stichotrich (*Sterkiella nova*) that has been stained to show its two macronuclei and four micronuclei. (b) A scanning electron micrograph of *Sterkiella nova*. Bars = 10 μm. Courtesy of Dr. K.G. Murti.

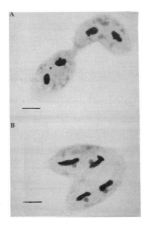

Figure 9.2 (a) A pair of stichotrichs in an early stage of mating (*Sterkiella histriomuscorum*). The two cells are connected by a cytoplasmic bridge. (b) A later stage of mating, in which the two cells are tightly joined. Bars — 10 μm.

(Figure 9.3a, b). Exchange of DNA now occurs by migration of a haploid micronucleus from each cell through the cytoplasmic channel into its partner (Figure 3c). The migratory haploid micronucleus fuses with a resident haploid micronucleus, forming a new diploid micronucleus in each cell (Figure 3d). The cytoplasmic channel is then resorbed, and the two cells go their independent ways. The new diploid micronucleus in a newly separated cell divides mitotically, without cell division (Figure 3e); one daughter micronucleus remains a micronucleus, and the other develops into a new macronucleus during the next several days. At the same time the old macronuclei and all the unused haploid micronuclei are destroyed. Finally, the newly formed macronucleus and the new

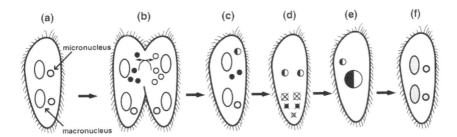

Figure 9.3 Steps in stichotrich mating. (a) A single organism before mating. (b) Two cells tightly joined and connected by a cytoplasmic bridge. The upper micronucleus has undergone meiosis, and the cells are about to exchange haploid micronuclei. (c) A postmating cell with a new, diploid micronucleus (in the top part of the cell). (d) The new, diploid micronucleus has divided by mitosis. The old macronuclei and the unused micronuclei are being destroyed. (e) One diploid micronucleus has developed into a new macronucleus. (f) The macronucleus and micronucleus have divided to reconstitute the appropriate nuclear numbers.

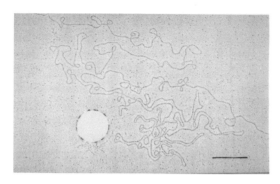

Figure 9.4 Electron micrograph of a small section of a micronuclear DNA molecule from *Sterkiella nova*. Bar = 1 μm. Courtesy of Dr. K. G. Murti.

diploid micronucleus divide without cell division to produce the appropriate number of each (e.g., two macronuclei and two micronuclei, Figure 3f).

The conversion of a micronuclear (germline) genome into a macronuclear (somatic) genome involves many remarkable manipulations of micronuclear DNA. Simultaneously, but not directly related to these DNA manipulations, the originally silent micronuclear genes are switched to their genetically active macronuclear form.

The complex processing of micronuclear DNA to create macronuclear DNA can be appreciated by comparing the structure of the two DNAs. Each of the approximately 100 micronuclear chromosome contains a single DNA molecule that is, on average, more than 1 million basepairs (bp) in length (Figure 9.4). The gene density is less than 5%, with gene-encoding regions, or genes, widely distributed along this DNA separated by long spacers of nongene DNA (Figure 9.5).

In sharp contrast, macronuclear DNA occurs in short molecules ranging in length from about 200 bp to 20,000 bp (Figure 9.6). There are about 25,000 different molecules (different in their nucleotide sequences); each molecule contains a single gene, or in a few cases 2 [14] or even 3 genes [11], yielding an estimate of 27,000 genes. The macronucleus contains none of the spacer

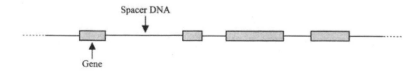

Figure 9.5 A diagram of the arrangement of genes and spacers in a small section of micronuclear DNA.

Figure 9.6 Electron micrograph of macronuclear DNA molecules from *Sterkiella nova*. Bar = 1 μm. Courtesy of Dr. K. G. Murti.

DNA that makes up most of the micronuclear DNA. Thus, in the conversion of a micronucleus to a macronucleus, the genes, representing less than 5% of the DNA, are excised from the micronuclear DNA, and all the spacer DNA, which makes up more than 95% of micronuclear DNA, is destroyed.

We can learn a little more about these events by microscopic observation. During the first 24 h (at 23°C) in the micronucleus-to-macronucleus conversion, polytene chromosomes are formed by 5 rounds, of replication of the DNA molecule in each chromosome in the diploid micronucleus. First, the two homologous copies of each chromosome align tightly with each other to form one composite chromosome. With each round of replication the copy number of DNA molecules in each of these composite chromosomes doubles. After five rounds, the resulting 64 identical copies in every composite chromosome remain aligned tightly in parallel to create a group of giant, or polytene chromosomes, as shown in the electron micrograph in Figure 9.7. Note that the polytene chromosomes have many cross-striations consisting of dark and light interbands. Each dark band results from a local condensation by tight coiling of a segment of each DNA molecule as it traverses the full length of a chromosome; in interbands the DNA molecules are not condensed, continuing in a straight form from band to band.

Once the polytene chromosomes are fully formed, they are destroyed in a very specific way that reflects major processing of the DNA molecules. A proteinaceous septum forms, cutting across each interband and transecting the 64 DNA molecules (Figure 9.8).

Figure 9.7 Electron micrograph of several polytene chromosomes isolated from a developing macronucleus of a stichotrich. Bar = 10 μm. Courtesy of Dr. K. G. Murti.

Transection of DNA has been confirmed by DNA purification at this stage; the very long micronuclear DNA molecules in the polytene chromosomes have been cut into shorter molecules, but they are still many times longer than the gene-sized molecules of a mature macronucleus. After transecting the interbands, the proteinaceous septa expand around each band, forming a string of vesicles (Figure 9.9).

If the developing macronucleus is broken open at this stage, the vesicles immediately disperse, each vesicle containing 64 copies of a short segment of the micronuclear DNA molecules. Most of each of these segments in a vesicle consists of spacer DNA, which is now destroyed, leaving short, gene-encoding molecules. Next, the vesicle walls dissolve, and the short DNA molecules go through (in *Sterkiella histriomuscorum*, for example) five rounds of replication, yielding a fully developed macronucleus with approximately 2000 copies

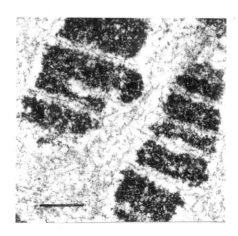

Figure 9.8 Electron micrograph of a section through a short segment of two polytene chromosomes showing transecting septa through interbands. Bar = 1 μm. Courtesy of Dr. K. G. Murti.

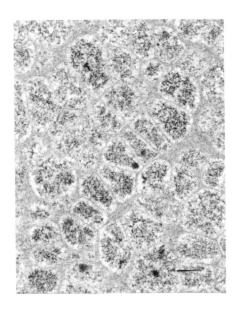

Figure 9.9 Electron micrograph
of vesicles formed from bands of a
polytene chromosome. Bar = 1 μm.
Courtesy of K. G. Murti.

of each gene-containing molecule. The new macronucleus and the new micronucleus divide, creating the mature complement of two macronuclei and two micronuclei and completing the 3-day process. At this point the cell resumes proliferation if food is available. The organism cannot mate again until it has undergone 20–30 cell divisions. The molecular basis of this sexually immature stage is unknown. If food is unavailable at the end of macronuclear development, the stichotrich may form a cyst. A cyst consists of a condensed, metabolically inert stichotrich inside a thick, stiff protective wall. In some species the cysts can be dried, frozen, and stored for many years. Brought to an appropriate temperature (e.g., 23°C) and supplied with food organisms, cysts hatch in 12–24 h, and the released ciliates resume proliferation.

There are major questions about the conversion of micronuclear DNA to macronuclear DNA for which we do not yet have satisfactory answers. A crucial issue is how the cell identifies the ends of gene segments in order to excise the genes from chromosomal DNA. The 50 bp at both ends of gene segments have an unusual nucleotide composition; one chain of the DNA is purine rich, and the other is pyrimidine rich [8], which conceivably serves as a signal for DNA cutting to excise the genes. However, nothing is known about the enzyme(s) that might recognize the 50 bp of unusual nucleotide composition and then carry out gene excision. Another important question is how the cell identifies spacer DNA for destruction and gene DNA for preservation. After gene segments are excised from micronuclear DNA, they are replicated to about 1000 copies each. We do not know how the cell is able to count the number

Figure 9.10 Diagram of the generalized structure of a macronuclear DNA molecule.

of replications so that the appropriate gene copy number is achieved. We do know that as each gene segment is released from chromosomal DNA, a repeat sequence of 36 nucleotides consisting of 3'GGGGTTTTGGGGTTTT 5', and so on, called a *telomeric repeat sequence*, is enzymatically added to the 3' end of both DNA chains. A complementary sequence of 20 bases, 5'CCCCAAAACC-CCAAAACCCC 3' is added to the 5' end of both chains, producing a telomeric structure with a 16-base single-stranded overhang, as shown in Figure 9.10. Special proteins known as *telomere binding proteins* bind to the single-stranded overhang, protecting the ends of the gene segments from attack by nucleases.

The elimination of the micronuclear spacer DNA removes the burden for the macronucleus of maintaining and replicating this large mass of nongenetic DNA that apparently makes no functional contribution to the cell. It permits the cell to carry an average of 1000 functional copies (in *Sterkiella histriomuscorum*) of each gene without creating an impractically large nucleus. Thus, the macronucleus represents the evolution of a nucleus with a gene density greater than 80% of the DNA and maximally streamlined for gene function, which means greatly increased RNA transcription. Increased transcription supports a higher rate of protein synthesis and of cell operations in general, which, in turn, promotes a faster rate of cell reproduction. Stichotrichs undergo cell division every 6–8 h when food organisms are plentiful. This is a high rate of reproduction for such very large cells. The strategy of a streamlined macronucleus likely accounts, at least in part, for the great success of stichotrichs in their ubiquitous and often dominant occupation of fresh and salt water and of soils throughout the world.

We have described phenomena surrounding the assembly of macronuclear genes from their encrypted micronuclear form. In the rest of this chapter we consider in more detail various aspects of the encryption process, the evolution of encryption, and the mechanism of gene assembly.

MICRONUCLEAR GENES ARE IN AN ENCRYPTED FORM

As striking a phenomenon as the creation of a genome with a very high gene density and with physically separate, macronuclear genes may be, it is not the most astonishing process that occurs during the conversion of a micronuclear genome to a macronuclear genome in stichotrichs.

Most genes in most organisms are interrupted by noncoding DNA sequences called introns. When a gene is transcribed into an RNA transcript, introns are transcribed as part of the transcript. Intron sequences are then spliced out of an RNA transcript to yield a messenger RNA molecule (mRNA) that is a faithful copy of the coding segments (exons) of the gene. The significance of introns is only partially understood; the transcripts of some genes can be processed in alternative ways so that different combinations of exons are spliced, creating different mRNA molecules that encode different protein products.

Stichotrich genes are intron-poor; only about 18% of the genes contain introns. However, their germline genes are interrupted by another kind of noncoding DNA called *internal eliminated segments*, or IESs. Unlike intron sequences, IESs are not transcribed, but are removed from DNA during conversion of micronuclear genes to macronuclear genes. Most IESs are short (less than 100 bp) and can be any sequence of A, T, G, and C bases, although they are particularly rich in A and T bases. IESs divide a gene into segments called *macronuclear destined segments*, or MDSs. Some micronuclear genes contain a single IES and therefore two MDSs. The gene encoding β telomere binding protein (βTP) in *Sterkiella nova* contains three IES of 32, 34, and 39 bp, respectively (Figure 9.11), dividing the micronuclear gene into four MDSs [9].

The micronuclear gene encoding βTP in another species, *Sterkiella histriomuscorum* is interrupted by six IESs, creating seven MDSs (Figure 8.11). One IES is in the 5' leader segment that precedes the coding region of the gene, and the other five interrupt the coding region. (The coding region is delimited by the start triplet codon, ATG, and the stop triplet codon, TGA.) Leader segments generally contain base sequences required to regulate transcription of the gene. Extrapolation from the IES numbers in 10 different micronuclear genes that

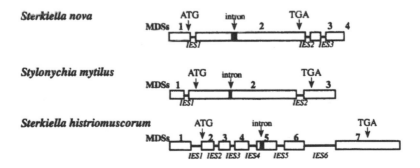

Figure 9.11 Diagram of the micronuclear βTP gene of *Sterkiella nova*, *Sterkiella histriomuscorum*, and *Stylonychia mytilus*. Macronuclear destined segments (MDSs) are open blocks connected by lines (internal eliminated segments). The intron is a filled (dark) block. ATG is the translation start codon, and TGA is the stop codon.

Figure 9.12 Diagram of the pair of repeats flanking internal eliminated segment (IES) 1 in the βTP gene of *Sterkiella histriomuscorum*. The macronuclear destined segments (MDSs) are in capital letters, and the IES is in lower case letters. The repeat sequences are underlined.

have been analyzed so far gives an estimate of at least 100,000 IESs, and therefore at least 100,000 MDSs, in one complete set (a haploid set) of micronuclear genes in *Sterkiella nova*. The presence of one or more IESs encrypts a gene so that it cannot be properly transcribed into a meaningful mRNA molecule. However, because micronuclear genes are inactive, useless transcription does not occur.

The vast number of IESs are excised from all the micronuclear genes during a few hours in the polytene chromosome stage of macronuclear development. Hence, the number of IESs that are excised is much greater than 100,000, perhaps as great as several million, depending on the number of DNA molecules in the polytene chromosomes. Correspondingly, 100,000 to several million MDSs must be precisely joined, as IESs are excised, to create incipient macronuclear genes competent for transcription.

An immediate question arises: how does the organism identify the junctions between MDSs and IESs, cut the DNA precisely at these junctions to remove the IESs, and ligate the MDSs? At least part of the answer lies in the nucleotide sequence at the ends of MDSs. For example, MDS 1 in the βTP gene of *Sterkiella histriomuscorum* diagrammed in Figure 9.11 ends with the nucleotide sequence 5' CAGTA 3' just before it joins IES 1, as shown in Figure 9.12. MDS 2 starts with the same five nucleotides where the end of IES 1 joins MDS 2.

Thus, the pair of direct repeat sequences of 5 bp may serve as a guiding signal for joining MDS 1 to MDS 2, simultaneously removing IES 1, as illustrated in Figure 9.13. A common, well-documented phenomenon in molecular genetics is the parallel alignment of identical DNA sequences, called *homologous pairing*; usually such paired sequences are present in two separate molecules (intermolecular pairing). In the case of the MDSs in Figure 9.13, we propose that the molecule forms a loop so as to align the two copies of CAGTA in parallel in the same molecule (intramolecular pairing). The aligned sequences are then recognized by an enzyme that cuts both repeats and rejoins them in a

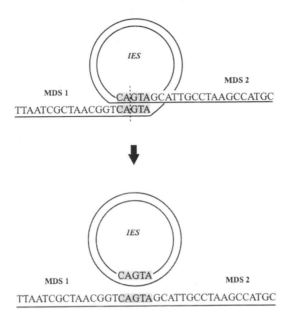

Figure 9.13 Diagram of alignment of the repeats in Figure 9.12 to create a loop of the internal eliminated segment (IES), which is removed by recombination.

switch configuration so that the IES, with one copy of the repeat, is removed as a circular molecule, and the two MDSs are joined through a copy of the repeat. Pairs of repeats are present without exception at the ends of MDSs that are destined to be joined. These pairs of repeat sequences flanking different IESs differ in length, ranging from 3 to 20 bp, and can be any sequence of the bases A, T, G, and C. Thus, to join the seven MDSs in the βTP gene and to clear it of IESs, six separate loops must be formed in the DNA of this micronuclear gene by aligning six pairs of repeat sequences.

ORIGIN OF IESs

To try to understand the significance of IESs, we have addressed questions of their origin and behavior during evolution. A proposed mechanism for the creation of IESs from spacer DNA is shown in Figure 9.14. The DNA of IESs consists of 75% to 100% A+T nucleotides. This is the same as the nucleotide composition of spacer DNA, but contrasts with the 48% to 68% A+T nucleotides in the coding region of stichotrich genes.

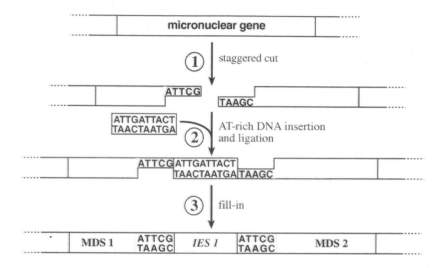

Figure 9.14 Model to explain the origin of an internal eliminated segment (IES) by insertion at a staggered cut (1) in micronuclear DNA of a short segment of AT-rich DNA (2). After insertion, filling the single-stranded gaps by complementary bases creates a pair of repeats in the flanking macronuclear destined segment (MDS) (3).

Segments of spacer DNA probably flood the cell transiently during macronuclear development when the polytene chromosomes are destroyed. A piece of blunt-ended spacer DNA is hypothesized to insert at staggered cuts in the double-stranded micronuclear DNA of a gene as shown in Figure 9.14. (Staggered cutting of DNA is a common phenomenon in molecular genetics.) After insertion, the single-stranded regions are filled in by DNA repair enzymes, creating a pair of direct repeats.

IES Migration

Because they are noncoding, mutations in IESs do not affect the organism. Thus, IESs are free to accumulate mutations, and they do so at a rate many times higher than do coding regions. These mutations include both changes in nucleotide sequence and shortening and lengthening of IESs. Mutations that change single bases are related to a particularly interesting behavior of IESs: IESs change their positions by migrating along a DNA molecule by the mutational mechanism illustrated by the IES in Figure 9.15.

1. In Figure 9.15, a mutation changes the first base in the IES from a to G (a is written in lowercase because it was a part of the IES). This

shortens the IES by one nucleotide, lengthens MDS 1 by 1 bp, and lengthens the repeat from CACG to CACGG.

2. A second mutation changes the first base in MDS 2 from C to t. This lengthens the IES by one base, shortens MDS 2 by one base, and shortens the repeat from CACGG to ACGG.

3. The third mutation changes the first base in the IES from t to A, which shortens the IES by two bases, lengthens MDS 1 by two bases, and changes the repeat from ACGG to ACGGAT.

4. A fourth mutation changes the second base in MDS 2 from C to t. This lengthens the right end of the IES by two bases and shortens MDS 2 by two bases. The repeat, which was originally CACG before the first mutation, is now very different (i.e., GGAT).

The overall effect of the four mutations, in addition to changing the sequence of the repeat, is migration of the IES to the right by three bases, adding GAT to the end of MDS 1 and subtracting CAC from the beginning of MDS 2. This migration has no effect on the coding properties of the MDSs. Excision of the IES and one copy of the repeat and ligation of the two MDSs at any step in Figure 9.15 yields a product with exactly the same sequence in the composite MDS. Thus, IES migration does not change the coding sequence of the gene, but it progressively changes the sequence and position of an IES, changes the sequence of the repeat pairs that flank it, and changes the sizes of the flanking MDSs.

IES migration may account for differences in IES positions in a particular micronuclear gene from one stichotrich species to another. The micronuclear βTP gene in *Sterkiella histriomuscorum* contains six IESs, as noted earlier in

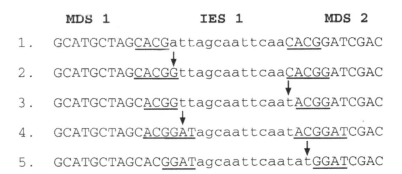

Figure 9.15 Internal eliminated segment (IES) migration by mutation of bases (see text for explanation).

Figure 9.11. The βTP gene in *Sterkiella nova* is interrupted by only three IESs and in *Stylonychia mytilus* by only two IESs (Figure 9.11). None of the IESs correspond in size, sequence, or position from one organism to another. For example, IES-1 (32 bp) in *Sterkiella nova* is in the gene-coding region (just after the ATG start codon). In *Stylonychia mytilus* IES 1 (20 bp) occurs in the leader region of the gene just before the start of the coding region. In *Sterkiella histriomuscorum* IES 1 (82 bp) is in the leader region but farther to the left of the start of the coding region than in *Stylonychia mytilus*. This could be interpreted to mean that IES 1 was inserted separately and independently in the three species during their divergence from a common evolutionary ancestor in which there was no IES. Alternatively, IES 1 may have been present in a common evolutionary ancestor, but mutations caused it to migrate to different positions and to change in sequence and size. Similarly, IES 2 in *Sterkiella nova* and *Stylonychia mytilus* and one of the IESs (e.g., IES 5 or 6) in *Sterkiella histriomuscorum* (see Figure 9.11) might have been derived from the same IES in a common ancestor. Whether or not this speculation is correct, it is clear that additional IESs have been inserted after the evolutionary divergence of three species, particularly in *Sterkiella histriomuscorum*, which has six IESs compared to two and three in the other two species.

Scrambling the Arrangement of MDSs by IES Recombination

IESs show yet another behavior during evolution that has a profound effect on micronuclear gene structure: IESs within the DNA molecule of a gene are able to recombine with one another. Such intramolecular recombination between IESs consists of breakage of two IESs and rejoining of broken ends in a switched configuration. For example, the micronuclear actin I gene of *Sterkiella nova* contains eight IESs and nine MDSs. Originally, eight IESs must have been inserted into the actin I gene in an ancestor of *Sterkiella nova*, creating nine MDSs probably in the orthodox order, 1-2-3-4-5-6-7-8-9. During evolution of *Sterkiella nova* from its ancestor, the nine MDSs have been rearranged into the scrambled order 3-4-6-5-7-9-2-1-8 by recombinations between IESs. An example of how IES recombination can generate MDS scrambling is shown in Figure 9.16.

In Figure 9.16a the nine MDSs in the micronuclear actin I gene in a theoretical ancestor of *Sterkiella nova* are arranged in their orthodox order. Figure 9.16b shows two IES recombinations: one recombination between IESs 4 and 5 and a second between IESs 4 and 6. For the recombinations to occur, the DNA must form two loops to bring IESs 4, 5, and 6 into parallel alignment. Recombination between IES 4 and 5 (indicated as an "x") creates the composite IES 4-5 that connects MDS 4 with MDS 6. A second recombination between IES 6 and IES 4 (indicated as an "x") creates the composite IES 4-6 connecting MDS 6 to MDS 5 and a third composite IES 5-4-6 connecting MDS 5 to MDS 7. As a result

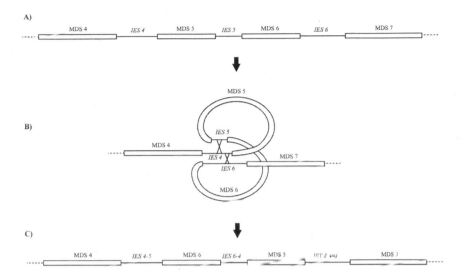

Figure 9.16 A model explaining how two internal eliminated segment (IES) recombinations change the orthodox order of macronuclear destined segment (MDS) 4-5-6-7 into the scrambled disorder, MDS 4-6-5-7 in the actin I gene of *Sterkiella nova*. (a) MDSs 4, 5, 6, and 7 in the orthodox order. (b) Folded micronuclear DNA and recombinations between IESs. (c) Scrambled arrangement of MDS created in b.

of the two recombinations, the orthodox MDS order, 4-5-6-7, has changed to the scrambled order, 4-6-5-7, which is the order observed for these four MDSs in the micronuclear actin I gene of *Sterkiella nova* (Figure 9.16c). Additional IES recombinations create the scrambled arrangement, MDSs 7-9-2-1-8. IES 1 (between MDSs 3 and 4) does not participate in recombination, so that the final scrambled arrangement is 3-4-6-5-7-9-2̄-1-8. MDS 2 in the micronuclear actin I gene of *Sterkiella nova* is a particular case of MDS scrambling because it is inverted, as indicated by the arrow over the 2. By convention the sequence of bases in a macronuclear gene is read from left to right in the 5' to 3' direction. This applies as well to the MDS in a nonscrambled micronuclear gene such as the seven MDSs in the βTP gene in Figure 9.11. In the actin I gene, eight of the nine MDSs have maintained their polarity (orientation); they are still read from left to right (5' to 3') in spite of being scrambled. MDS 2 is in the reversed (inverted) polarity (3' to 5') as a result of recombination between the two IESs that flank it. How MDS 2 became inverted is postulated in Figure 9.17. It is assumed that the micronuclear actin I gene first became scrambled into the MDS arrangement, 3-4-6-5-7-9-2-1-8, with MDS 2 in the same (noninverted) polarity as the other eight MDSs (Figure 9.17a). Folding the molecule into a

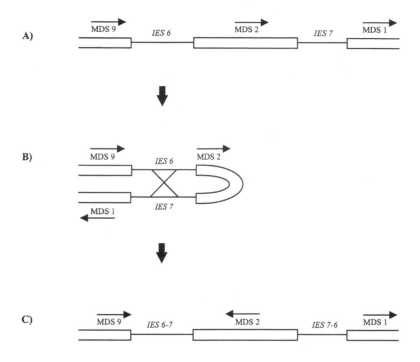

Figure 9.17 A model explaining how internal eliminated segment (IES) recombination brought about inversion of macronuclear destined segment (MDS) 2 in the actin I gene of *Sterkiella nova* during evolution. (a) Original arrangement of MDSs 9, 2, and 1. (b) DNA folded into a hairpin, and recombination between IES 6 and 7. (c) Inversion of MDS 2 as a result of the IES recombination.

hairpin allows the flanking IESs 6 and 7 to align with each other (Figure 9.17b). Recombination between the two IESs inverts MDS 2 so that it now reads right to left (Figure 9.17c; i.e., it is inverted).

Evolution of IES Additions and MDS Scrambling in the Actin I Gene

Examining the actin I gene in a series of nine stichotrich species shows how IESs have been added to the micronuclear actin I gene and how the MDSs became scrambled during evolution. First, however, we can arrange these nine organisms in an evolutionary tree based on the degree of similarity of the base sequences in their genes encoding ribosomal RNA (called the rDNA gene) (Figure 9.18).

The longer ago any two organisms diverged from each other in evolution, the more time has been available for mutations to accumulate and the less their

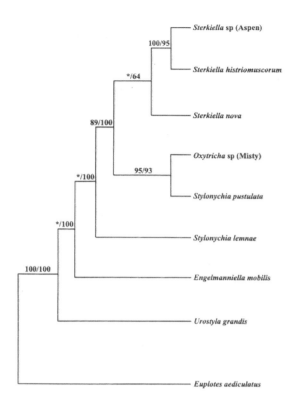

Figure 9.18 Phylogenetic relationships among nine different ciliates based on sequences of the rDNA gene (Hewitt et al., unpublished).

rDNA sequences match. Thus, in comparisons between pairs of organisms, the sequence of the rDNA gene in *Euplotes aediculatus* has the least similarity with the rDNA sequence of any of the other eight organisms. Consequently, *Euplotes* must have been the first to split away from the ancestral line leading to the other eight organisms in Figure 9.18. *Urostyla grandis* split away next, followed by *Engelmanniella mobilis*, and then by *Stylonychia lemnae*. Finally, the evolutionary line split into two lines, one leading to the three *Sterkiella* species and the other to *Stylonychia pustulata* and *Oxytricha* sp. (Misty).

In Figure 9.19 the structures of the micronuclear actin I genes in the nine organisms are arranged in the evolutionary pathway defined by the rDNA tree in Figure 9.18. The pathway begins with a common progenitor of the nine organisms, whose micronuclear actin I gene contained no IESs. The common progenitor gave rise to two lines of descent (Figure 9.19). One line led to contemporary *Euplotes aediculatus*, in which no IESs have been inserted into

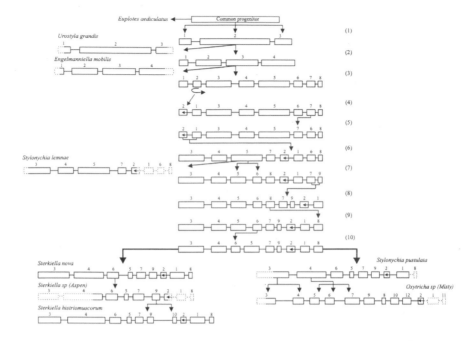

Figure 9.19 A scheme describing the origin and evolution of internal eliminated segments (IESs), macronuclear destined segments (MDSs), and MDS scrambling in the micronuclear actin I gene in nine species of hypotrichs. The scheme begins with a common ancestor, which has no IESs in the micronuclear actin I gene. *Euplotes* derived from this ancestor without addition of any IESs. IES additions and recombinations create an increasingly complex structure of the actin I gene during evolution of the other eight organisms. The scheme ends in two branches, one branch ending with three *Sterkiella* species and the other branch with *Stylonychia pustulata* and *Oxytricha* sp. (Misty).

the actin I gene; in the other line, IESs were progressively added in stepwise fashion.

 After two IESs had been added [1 in Figure 9.19], the line split, giving rise to *Urostyla grandis*, which currently has only two IESs and three nonscrambled MDSs and a line in which a third IES was inserted [2 in Figure 9.19]. The stichotrich *Engelmanniella mobilis*, with its three IESs and four nonscrambled MDSs, diverged at this point. Evolution of the actin I gene continued with the addition of four more IESs, creating a total of eight MDSs [3 in Figure 9.19]. A series of IES recombinations inverted MDS 2 and scrambled the eight MDSs into the order, 3-4-5-7-$\overset{\leftarrow}{2}$-1-6-8 [4, 5, and 6 in Figure 9.19]. *Stylonychia lemnae*, which has retained this order of eight MDSs, diverged from the pathway at this point. The pathway continues, leading to the remaining stichotrichs in Figure 9.19. Addition of another IES [7 in Figure 9.19] and further IES recombinations [8, 9, and 10 in Figure 9.19] produced the MDS pattern, 3-4-6-5-7-9-$\overleftarrow{2}$-1-8. The pathway then split. In one branch the pattern was inherited by *Sterkiella nova*, which, in turn, gave rise to *Sterkiella* sp. (Aspen) and, by addition of another IES, to *Sterkiella histriomuscorum*. The other branch evolved first into *Stylonychia pustulata* and, finally, by addition of three more IESs to the 3-4-6-5-7-9-$\overleftarrow{2}$-1-8 pattern into *Oxytricha* sp. (Misty) with 12 MDSs, 3-4-5-6-7-9-8-10-12-$\overleftarrow{2}$-1-11. IES additions and MDS scrambling in Figure 9.19 not only conform to the pathway defined by rDNA sequences, but it is also the most parsimonious series of steps explaining the evolution of actin I gene structure. This evolutionary scheme in Figure 9.19 may be further refined when the structure of the micronuclear actin I gene is determined in additional stichotrichs. For example, some of the theoretical intermediates might be verified, but it is unlikely that the overall evolutionary pattern will require any major rearrangements. Thus, we have compelling evidence that IESs are added sequentially in evolution and that these IES additions are intermixed with sporadic IES recombinations that scramble MDSs.

ASSEMBLY OF MACRONUCLEAR GENES

We noted earlier that the pair of repeat sequences in the ends of MDSs immediately flanking an IES in the micronuclear βTP gene offers a clue about how a stichotrich ciliate removes the IES during macronuclear development (see Figure 9.13). Recombination between the repeats excises the IES, along with one copy of the repeat, and ligates the two MDSs into a composite MDS. In some micronuclear genes, recombination between IESs during evolution has rearranged MDSs into a scrambled order, in some instances inverting MDSs. In this case, too, the pairs of repeat sequences are crucial for excision of IESs as well as for ligation of MDSs into the orthodox order, including reinversion of MDSs into their 5' → 3' orthodox polarity. We present now three molecular operations that accomplish the assembly into its macronuclear form of any micronuclear gene composed of nonscrambled or scrambled MDSs or a combination of nonscrambled and scrambled MDSs.

First, we refer to the pairs of repeats in the ends of MDSs as a pair of pointers that points one MDS to join with another MDS in the appropriate order during gene assembly. An MDS has an incoming pointer and an outgoing pointer, as diagrammed in Figure 9.20 for the four MDSs in the micronuclear βTP gene of *Engelmanniella mobilis*. For example, MDS 2 has an incoming pointer (P2) in its 5' end (left end) and an outgoing pointer (P3) in its 3' end (right end). A sequence can act as a pointer only if it is at the boundary between an MDS and IES.

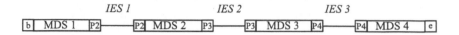

Figure 19.20 Diagram of the micronuclear βTP gene in *Sterkiella nova*. Pointers at the ends of macronuclear destined elements (MDSs) are designated by P. An MDS has an outgoing pointer and an incoming pointer except terminal MDSs (i.e., MDSs 1 and 4). MDS 1 begins with b for "beginning," and MDS 4 ends with e for "end." The b and e are sites to which telomere sequences are added when the gene is finally excised from its chromosome.

The first of the three operations by which IESs are removed and MDSs are ligated we call *loop, direct repeat excision*, or *ld excision* for short. This operation deals with any two MDSs that are in the orthodox order such as any two MDSs in the micronuclear βTP gene or MDSs 3 and 4 in the actin gene of *Sterkiella nova* (Figure 9.19). An ld excision applied to IES 2 in the βTP gene in Figure 9.20 is illustrated in Figure 9.21.

The second operation, called *hairpin, inverted repeat-excision*, or *hi excision*, applies to MDSs that are inverted, and therefore the two pointers in a pair are inverted relative to one another. For example, assume that the order of MDSs in a micronuclear gene is 1-2-$\overleftarrow{3}$-4 (the gene is shown in Figure 9.22). Because MDS 3 is inverted, its incoming pointer P3 and its outgoing pointer P4 are also necessarily inverted. Folding of the molecule into a hairpin in Figure 9.22 reinverts MDS 3 and allows the P3 pointer in MDS 2 to align with the P3 pointer in MDS 3. Recombination between the two P3 pointers then joins MDSs 2 and 3 in a composite MDS with orthodox polarity throughout and shifts the inverted IES 2 to a position adjacent to the recombined composite MDS. MDS 3 was reinverted above by only one application of hi excision. This was possible because its partner for applying hi excision, MDS 2, is separated from it only by an IES. More commonly, two MDSs that engage in hi excision are separated by more than just an IES. Therefore, more than one application of hi excision may be required interspersed by applications of other operations.

The third operation is called *double loop, alternating direct repeat-excision*, or *dlad excision*. It applies to a molecule with two pairs of pointers in which the segment of a DNA molecule encompassed by one pair of pointers overlaps with the segment enclosed by the second pair of pointers.

This is illustrated in Figure 9.23, in which the DNA segment encompassed by the two copies of P4 overlaps with the segment encompassed by the two copies of P2 (i.e., one of the copies of P2 resides between the two copies of P4). To resolve this scrambled state (MDS 1-3-2-4), the molecule folds into

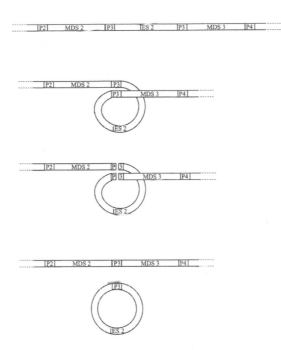

Figure 19.21 Diagram of ld excision of IES 2 in the βTP gene in Figure 9.20. The molecule folds in such a way that pointers P3 become aligned. Recombination between pointers P3 excises the internal eliminated segment (IES) together with a hybrid copy of P3 and ligates macronuclear destined segment (MDSs) 2 and 3 through a hybrid copy of P3.

two loops so that the two copies of P2 are aligned in one loop, and the two copies of P4 are aligned in the other loop. Thus, two recombinations take place: recombination between the two copies of P2 joins MDS 1 to MDS 2 into a composite MDS through a hybrid copy of P2, and recombination between the two copies of P4 joins MDS 3 to MDS 4 into another composite MDS through a hybrid copy of P4. As a result of these two recombinations, IESs 3, 2, and 1 are joined together by the other hybrid copies of P2 and P4. This composite IES is now interposed between MDSs 2 and 3 and may be removed by a subsequent operation of ld excision.

The three molecular operations presented above are *intramolecular*—in each of these operations a molecule reacts with (folds on) itself (i.e., it folds on itself, and recombination takes place between some segments within the molecule). The gene assembly process using the three operations discussed above is presented in detail elsewhere [2, 3, 10, 13], including the computational properties of the process. The computational aspects of gene assembly are also discussed

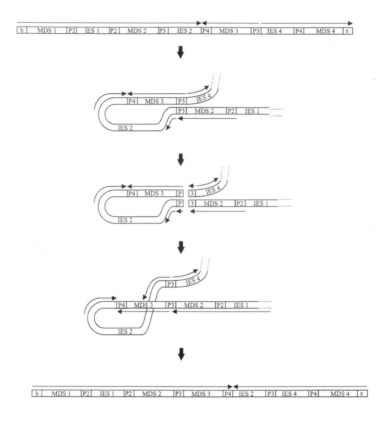

Figure 9.22 The operation hi excision reinverts inverted macronuclear destined segment (MDS) 2 in the actin 1 gene of *Sterkiella nova*. The molecule folds in a hairpin so as to align the pair of pointers P2 in the same polarity (indicated by arrows). Recombination between P2 yields a molecule with MDS 1 joined to MDS 2 with MDS 2 reinverted into the same polarity as MDS 1.

in a series of papers by Kari and Landweber [5]; however, their model is intermolecular, meaning that they allow different molecules to recombine.

Assembly of the Macronuclear Actin I Gene

In the micronuclear actin I gene of *Urostyla grandis* and *Engelmanniella mobilis* the MDSs are in the orthodox order. Therefore, assembly of the macronuclear version of the gene can proceed by the ld excision operation (i.e., formation of single loops that align pairs of repeats), followed by recombination between the repeats in a pair, as described earlier for the micronuclear βTP gene. Assembly of the macronuclear actin I gene with its scrambled MDSs (in six species in Figure 9.19) is more complex because the MDSs must be rearranged into the

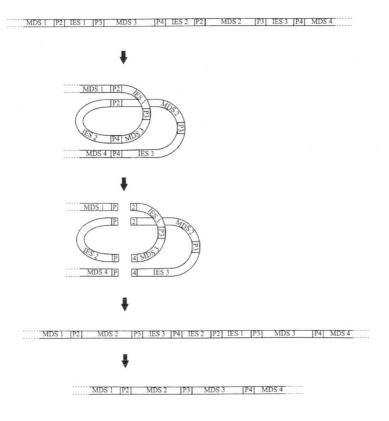

Figure 9.23 The dlad excision operation unscrambles macronuclear destined segments (MDSs) in which a segment of DNA encompassed by one pair of pointers (in this case P2) overlaps with a DNA segment encompassed by another pair of pointers (P4). The molecule folds into a double loop that allows pointer pairs P2 and P4 to align. Recombination between pointers joins MDS 1 to MDS 2 and MDS 3 to MDS 4. A subsequent ld excision operation removes the composite internal elimi- nated segment (IES) 3-2-1 and joins MDS 2 with MDS 3.

orthodox order. The order of MDSs in the micronuclear actin I gene in *Sterkiella nova* is 3-4-6-5-7-9-$\overset{\leftarrow}{2}$-1-8. Table 9.1 gives the incoming and outgoing pointers for each of the MDSs, which are listed in their order in the scrambled gene.

The pointers shown in Table 9.1 identify how MDSs are to be joined in the actin I gene of *Sterkiella nova*. Thus, for example, MDS 7 is separated from MDS 8 by MDSs 9-$\overset{\leftarrow}{2}$-1 and four IESs. The right end (the 3' end) of MDS 7 contains the outgoing pointer P8, CTTTGGGTTGA, and the left end of MDS 8 (the 5' end) to which it must join contains the incoming pointer, P8. The following succession of operations will assemble the micronuclear actin I gene

Table 9.1 The eight pairs of pointers in the micronuclear actin I gene in *Sterkiella nova*.

Pointers in 5' of MDS	MDS	Pointers in 3' of MDS
GGAGTCGTCAAG (P3)	3	AATC (P4)
AATC (P4)	4	CTCCCAAGTCCAT (P5)
GCCAGCCCC (P6)	6	CAAAACTCTA (P7)
CTCCCAAGTCCAT (P5)	5	GCCAGCCCC (P6)
CAAAACTCTA (P7)	7	CTTTGGGTTGA (P8)
AGGTTGAATGA (P9)	9	3 TAS
CTTACTACACAT (P2)	2	GGAGTCGTCAAG (P3)
5 TAS	1	CTTACTACACAT (P2)
CTTTGGGTTGA (P8)	8	AGGTTGAATGA (P9)

into its macronuclear form: two ld excisions, a dlad excision, an ld excision, a dlad excision, and, finally, two hi excisions [10].

Assembly of the Macronuclear Molecule Encoding α Telomere Binding Protein

In the actin I gene the MDSs appear to be scrambled in no particular order. In the micronuclear gene encoding α telomere binding protein (αTP) in three stichotrich species, the MDSs are scrambled in a predominantly odd/even pattern, with no inverted MDSs (Figure 9.24). *Sterkiella nova* and *Stylonychia mytilus* have identical patterns with the MDS order: 1-3-5-7-9-11-2-4-6-8-10-12-13-14, although the corresponding IESs in the two organisms are different in size and sequence and have migrated, which changes the sizes of MDSs [12].

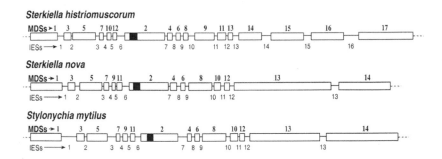

Figure 9.24 Diagrams of the micronuclear αTP gene in three stichotrichs. Macronuclear destined elements (MDSs) are open blocks, internal eliminated segments (IESs) are lines between blocks, and introns are dark blocks (in MDS 2). From Prescott et al. [12].

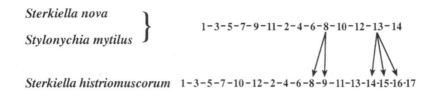

Sterkiella nova

Stylonychia mytilus } 1-3-5-7-9-11-2-4-6-8-10-12-13-14

Sterkiella histriomuscorum   1-3-5-7-10-12-2-4-6-8-9-11-13-14-15-16-17

Figure 9.25 Evolution of the micronuclear αTP gene of *Sterkiella histriomuscorum* from the αTP gene of *Sterkiella nova/Stylonychia mytilus* by insertion of three more internal eliminated segments, creating three more macronuclear destined segments. From Prescott et al. [12].

The gene in the third species, *Sterkiella histriomuscorum*, contains three additional IESs and MDSs. MDS 13 in *Sterkiella nova* and *Stylonychia mytilus* is split into three MDSs in *Sterkiella histriomuscorum* by insertion of two IESs, and MDS 8 has been split into two MDSs by insertion of an IES, as illustrated in Figure 9.25.

A diagram of 14 MDSs, with their incoming and outgoing pointers, in the micronuclear αTP gene of *Sterkiella nova* or *Stylonychia mytilus* is shown in Figure 9.26. Eleven of the 13 pairs of pointers encompass overlapping segments of DNA (e.g., the pair of P2 pointers encompasses a segment of DNA that overlaps with DNA segments encompassed by the pointer pairs P3, P4, P5, P6, P7, P8, P9, P10, P11, and P12). Thus, a series of six dlad excision operations will rearrange MDSs 1 through 12 into the orthodox order. Two ld excision operations join MDSs 12, 13, and 14, completing assembly of the αTP gene in *Sterkiella nova* and *Stylonychia mytilus*. The αTP gene in *Sterkiella*

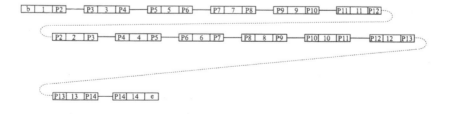

Figure 9.26 A diagram of the 14 macronuclear destined segments (MDSs), with their pointers, of the micronuclear αTP gene in *Sterkiella nova* and *Stylonychia mytilus*. MDSs are numbered in the open blocks. Internal eliminated segments (not numbered) are lines connecting MDS blocks.

histriomuscorum contains three more MDSs (17 total), which simply require an additional three ld excision operations to assemble the macronuclear gene.

Assembly of the Macronuclear Gene Encoding DNA Polymerase α

Ten micronuclear genes have been examined in several stichotrichs, and three are scrambled. The third scrambled gene, the DNA *polα* gene, has the most complex MDS scrambling pattern. In *Sterkiella nova* the micronuclear gene contains 45 MDSs, most of which are arranged in an odd/even pattern, as in the αTP gene, but with inversion of 24 of the MDSs relative to the other 21 MDSs (Figure 9.27). In addition, MDSs 29-31-33-35-37-39-41-43 and the seven IESs between these MDSs form a separate group at a different, unknown location in the genome.

In addition to the DNA *polα* gene in *Sterkiella nova*, the structure of the micronuclear DNA *polα* gene has been determined in *Sterkiella histriomuscorum* and *Stylonychia lemnae*. The patterns of scrambling are similar, although IESs have changed in size and sequence after divergence of the three species from one another. Migration of IESs has caused modest changes in the sizes of homologous MDSs, and the MDS number has increased to 48 in *Stylonychia lemnae* and to 51 in *Sterkiella histriomuscorum* by insertion of three and six IESs, respectively. Assembly of this complicated gene structure into its

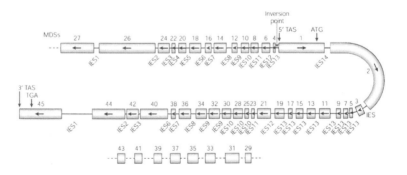

Figure 9.27 A diagram of the micronuclear DNA *polα* gene of *Sterkiella nova*. The gene occurs in two groups of macronuclear destined segments (MDSs) in different locations in the chromosomal DNA. Because of its length, the larger group is bent into two lines. It contains an inversion indicated by the long vertical arrow above the top line of MDSs. Direction of reading of the MDSs is indicated by horizontal arrows in the MDS blocks. Internal eliminated segments (IESs) are lines between blocks. 5' TAS and 3' TAS (telomere addition sites) indicate where telomere sequences are added when the gene is excised from the chromosome. ATG is the start codon for translation, and TGA is the stop codon. From Prescott and DuBois [9].

functional, macronuclear form in *Sterkiella histriomuscorum* requires all three operations.

Creation of the Odd/Even Pattern of MDSs in the αTP and the DNA *polα* Gene

The regular odd/even pattern of MDSs in the αTP and DNA *polα* genes implies a coordinated formation of MDSs by insertion of IESs during evolution. A model describing coordinated formation of MDSs in the micronuclear αTP gene in *Sterkiella nova* or *Stylonychia mytilus* is presented in Figure 9.28 [12].

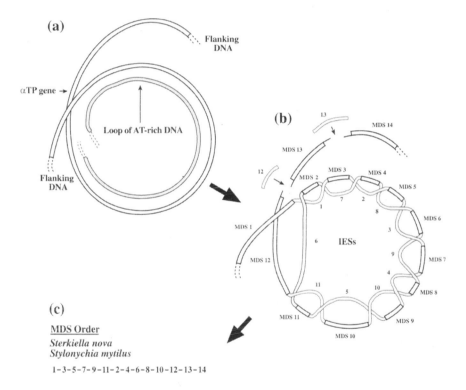

Figure 9.28 A model to explain the creation of the odd/even pattern of macronuclear destined segments (MDSs) in the αTP gene. (a) A loop of AT-rich DNA aligns with the original αTP gene that has no internal eliminated segments (IESs). (b) A series of recombinations between the loop of AT-rich DNA and the αTP gene creates an odd/even pattern of MDS 1–12. MDSs 13 and 14 are formed by two separate, independent insertions of IESs. (c) The final order of MDSs in the αTP gene. Redrawn from Prescott et al. [12].

The model the αTP gene, which initially lacks IESs, aligns with a loop of AT-rich spacer DNA (Figure 9.28). A series of recombinations between the gene and the AT-rich loop creates MDSs 1–12, connecting MDSs 1, 3, 5, 7, 9, and 11 in one series, and MDSs 2, 4, 6, 8, 10, and 12 in another series. A recombination between two segments of the AT-rich loop creates IES 6, connecting MDS 11 with MDS 2. Nonscrambled MDSs 12, 13, and 14 are created as separate events by insertion of two IESs either before or after the main set of recombinations. The model can be applied to the DNA *polα* gene as well.

TEMPLATE-GUIDED MODEL OF RECOMBINATION IN GENE ASSEMBLY

The pointers in MDSs that flank IESs range in length from 3 to 8 basepairs and average 4 basepairs in the case of those IESs separating two numerically consecutive MDSs. These include not only IESs in nonscrambled genes but also some IESs in genes in which some of the MDSs are scrambled. For example, IES 1 in the actin I gene of *Sterkiella nova* is flanked by MDSs 3 and 4 (i.e., they are nonscrambled relative to each other; the incoming and outgoing pointers P4 in these MDSs is 4 bp. However, the pointers in MDSs that are scrambled relative to each other range from 7 to 19 bp, with an average of 11 bp. The reason for the longer pointers in scrambled MDSs is surely significant, but it is not understood.

Clearly, in the case of nonscrambled MDSs, the pointers are too short to specify unambiguously the correct alignment of the pointers for recombinational excision of the IESs. A pointer of 4 bp (the average length for nonscrambled MDSs) will occur by chance on average every 4^4, or 256 bp. Thus, the probability is high that a copy of a 4-bp pointer will occur within an IES itself or in one or both MDSs within a few hundred basepairs of the IES. How are the correct copies of the pointer chosen for alignment and recombination? A specific example of the problem is the pair of 3-bp pointers P7 that flank IES 6 (separating nonscrambled MDSs 6 and 7) in the actin I gene of *Sterkiella histriomuscorum* (see Figure 9.19). This 3-bp pointer sequence (AGT) occurs five times within the IES, again in MDS 7, 8 bp downstream of the IES, and in MDS 6, just 44 bp upstream of the IES, creating many possibilities for misidentification of pointers. Misidentification of the correct copies of pointers in the case of scrambled MDSs might seem to be much less of a problem because the pointers are longer. The shortest pointer, 6 bp, will occur by chance on average once every 4096 bp. However, members of some pointer pairs must be brought into alignment from distances much greater than 4096 bp (e.g., the pointers in the eight MDSs of the DNA *polα* gene that are in a completely different location in the genome), leaving open the possibility for incorrect identification of the correct copies of pointers.

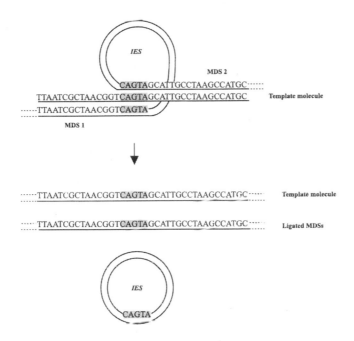

Figure 9.29 General view of the template-guided model of gene assembly in stichotrichs. A DNA molecule from the old macronucleus aligns macronuclear destined segments (MDSs) in the orthodox order by homologous pairing with the MDSs. Such alignment brings the correct pointers (CAGTA) into position for recombination in an ld excision operation. The nucleotide sequences are shown for the 5' → 3' chain of DNA double helices. The complementary chains are shown as lines. The excised internal eliminated segment (IES) is shown as a circle, but the excised IES could be linear, depending on how the recombination is postulated to occur.

We propose a template-guided model of gene assembly to assure alignment of correct copies of pointers. The model is based on work with *Paramecium* in the laboratories of Meyer [1] and Forney [4] that shows that the old macronucleus can guide IES excision in a new, developing macronucleus. The essence of the template model for gene assembly in stichotrichs, which is described in detail in Prescott, Ehrenfeucht, and Rozenberg (submitted), is shown in Figure 9.29.

In the model, copies of DNA molecules from the old macronucleus enter the developing macronucleus and align by homologous pairing to MDS sequences in the DNA of the polytene chromosomes. In Figure 9.29 alignment of an old macronuclear sequence with two numerically consecutive MDSs in the DNA in polytene chromosomes aligns the pointers in those MDSs, producing a looping out of the IES. Recombination between members of the pointer pair excises the IES and joins the MDSs through one copy of the pointer, in this case by application of the ld excision operation. The template molecule emerges intact and can be used repeatedly to guide more recombinations in the multiple DNA molecules in a polytene chromosome.

TEMPORAL ORDER OF EVENTS IN GENE ASSEMBLY

There are at least three possible models, distinguished by their temporal programs, for IES excisions and MDS ligations during macronuclear development: (1) all IESs might be excised at the same time, with simultaneous joining of all MDSs; (2) IESs might be excised and MDSs ligated in a fixed temporal order, creating one or more partially processed intermediates; (3) IESs might be excised and MDSs ligated in an order that is not fixed, with different orders in different developing macronuclei or even with different orders in different DNA molecules within the same developing macronucleus. One experimental result so far suggests IESs are excised and MDSs are ligated in a fixed temporal order. IES 6 in the actin I gene of *Sterkiella histriomuscorum*, which separates MDS 9 from MDS 10, is invariably the first to be excised, followed some hours later by excision of the remaining eight IESs and MDS reordering/ligation. Whether these remaining events occur in a fixed temporal order has not yet been determined.

CONCLUSIONS

In this chapter we described in detail the decryption of micronuclear genes in ciliates. We considered the gene assembly process from a computational perspective and introduced the template-guided model. The temporal order of events in gene assembly was also briefly considered, and we highlighted some open questions on this matter. In the next chapter, Kari and Landweber further develop the decryption of macronuclear DNA from a computational standpoint.

Acknowledgments   This work is supported by NIGMS research grant R01 GM 56161 and NSF research grant MCB-9974680 to D.M. Prescott. We are grateful to Gayle Prescott for her patience in typing several versions of the manuscript and to Beth Hewitt for constructing the figures.

References

[1] S. Duharcourt, A-M. Keller, and E. Meyer. Homology-dependent maternal inhibition of developmental excision of internal eliminated sequences in *Paramecium tetraurelia. Mol. Cell. Biol.*, 18:7075–7085, 1998.

[2] A. Ehrenfeucht, T. Harju, I. Petre, D. M. Prescott, and G. Rozenberg. Formal systems for gene assembly in ciliates. *Theor. Computer Sci.*, 292(1):199–219, 2003.

[3] A. Ehrenfeucht, I. Petre, D. M. Prescott, and G. Rozenberg. String and graph reduction systems for gene assembly in ciliates. *Math. Struct. Computer Sci.*, 12(2):113–134, 2002.

[4] J. D. Forney, F. Yantiri, and K. Mikami. Developmentally controlled rearrangement of surface protein genes in *Paramecium tetaurelia*. *J. Eukar. Microbiol.*, 43:462–467, 1996.

[5] L. Kari and L. F. Landweber. Computational power of gene rearrangement. In E. Winfree and D. Gifford, editors, *Proceedings of DNA Based Computers V*, pages 207–216. American Mathematical Society, Providence, RI, 1999.

[6] D. M. Prescott. The DNA of ciliated protozoa. *Microbiol. Rev.*, 58:233–267, 1994.

[7] D. M. Prescott. Genome gymnastics: unique modes of DNA evolution and processing in hypotrich ciliates. *Nature Rev. Genet.*, 1:191–198, 2000.

[8] D. M. Prescott and S. J. Dizick. A unique pattern of intrastrand anomalies in base composition of the DNA in hypotrichs. *Nucl. Acids Res.*, 28:4679–4688, 2000.

[9] D. M. Prescott and M. L. DuBois. Internal eliminated segments (IESs) of *Oxytrichidae*. *J. Eukar. Microbiol.*, 43:432–441, 1996.

[10] D. M. Prescott, A. Ehrenfeucht, and G. Rozenberg. Molecular operations for DNA processing in hypotrichous ciliates. *J. Protistol*, 37:241–260, 2001.

[11] D. M. Prescott, J. D. Prescott, and R. M. Prescott. Coding properties of macronuclear DNA molecules in *Sterkiella nova* (*Oxytricha nova*). *Protist*, 153(1):71–78, 2002.

[12] J. D. Prescott, M. L. DuBois, and D. M. Prescott. Evolution of the scrambled germline gene encoding α-telomere binding protein in three hypotrichous ciliates. *Chromosoma*, 107:293–303, 1998.

[13] G. Rozenberg. Gene assembly in ciliates: computing by folding and recombination. In A. Salomaa, D. Wood, and S. Yu, editors, *A Half-Century of Automata Theory: Celebration and Inspiration*, pages 93–103. World Scientific, NJ, London, 2001.

[14] A. Seegmiller, K. R. Williams, and G. Herrick. Two two-gene macronuclear chromosomes of the hypotrichous ciliates *Oxytricha fallax* and *O. trifallax* generated by alternative processing of the 81 locus. *Devel. Genet.*, 20:348–357, 1997.

10

Biocomputation in Ciliates

Lila Kari and Laura F. Landweber

Ciliates are unicellular protists that may have arisen more than a billion years ago. They have since diverged into thousands of species, many uncharacterized, the genetic divergence among ciliates being at least as deep as that between plants and animals [17]. Despite their diversity, ciliates are united by two common features; the presence of short threads called cilia on their surface, whose rhythmic beating causes movement and is also useful for food capture, and the presence of two types of nuclei. The *macronucleus* contains DNA encoding functional copies of all the genes that regulate vegetative growth and cell proliferation. The *micronucleus* contains encrypted versions of the macronuclear DNA, is mostly functionally inert, and is only used for sexual exchange of DNA. In this chapter we study the decryption of the macronuclear DNA from a computational perspective.

INTRODUCTION

When two cells mate, they exchange micronuclear information. After they separate, the old micronuclei and macronuclei degenerate, while the newly formed micronuclei develop into new macronuclei over hours or days, depending on the species. Few ciliates have so far been studied at the level of molecular genetics: *Tetrahymena* and *Paramecium* representing the *Oligohymenophorans* and *Oxytricha* (recently renamed *Sterkiella*), and *Stylonichia* and *Euplotes* representing *Spirotrichs*. The DNA molecule in each of the approximately 120 chromosomes in the micronucleus contains on average approximately 10^7 base-pairs (bp) in *Oxytricha* species and approximately 18×10^6 bp in *Stylonichia lemnae* [17]. The size of the DNA molecules in the macronucleus is, in contrast, very small. In various *Oxytricha* species and *S. lemnae*, macronuclear DNA molecules range in size from 400 to 15,000 bp with most molecules in the 1000–8000 bp range [18].

Macronuclear DNA sequences are derived from the micronuclear sequences through a series of DNA rearrangements as follows. The segments that together constitute a macronuclear sequence (*macronuclear destined sequences* or MDSs) are present as sub-sequences in the micronuclear DNA. However, in the micronuclear DNA, MDSs are interspersed with long DNA sequences (*internal eliminated sequences* or IESs) that are excised in the micronucleus to macronucleus differentiation. (Note that excision of IESs from micronuclear DNA is distinct from excision of introns, which occurs at the RNA level after transcription of macronuclear DNA.) IESs and intergenic DNA represent large regions of the micronuclear DNA. In *Oxytricha* species, only 4% of the micronuclear DNA represents macronuclear destined sequences, while in *S. lemnae* that proportion is still smaller, 2% [18].

In addition, in some spirotrich micronuclear genes, the MDSs are present in a permuted order in the micronuclear DNA, relative to the "correct" macronuclear order [18], as shown in Figure 10.1. For example, the micronuclear actin I gene in *Oxytricha nova* consists of nine MDSs separated by eight IESs. The order of MDSs in the micronucleus is 3-4-6-5-7-9-2-1-8 [19].

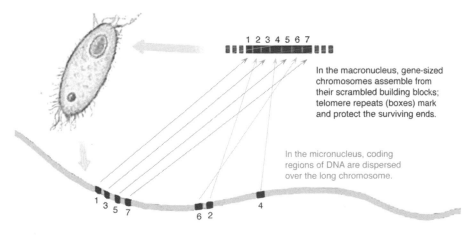

1 2 3 4 5 6 7

In the macronucleus, gene-sized chromosomes assemble from their scrambled building blocks; telomere repeats (boxes) mark and protect the surviving ends.

In the micronucleus, coding regions of DNA are dispersed over the long chromosome.

1 3 5 7   6 2   4

Figure 10.1 Overview of gene unscrambling. Dispersed coding macronuclear destined segments 1–7 reassemble during macronuclear development to form the functional gene copy (top), complete with telomere addition to mark and protect both ends of the gene. From Landweber and Kari [8].

The gene encoding the α telomere binding protein in *Oxytricha nova* is present in the micronucleus in the permuted order 1-3-5-7-9-11-2-4-6-8-10-12-13-14 [12]. The gene encoding DNA polymerase α in *S. lemnae* is apparently broken into more than 48 MDSs scrambled in an odd/even order [10], with 14 MDSs inverted on the opposite strand of another 24 MDSs, 9 additional MDSs on a separate locus, and 1 MDS not present on either locus. MDSs have to be reordered and intervening IESs eliminated to construct the functional macronuclear gene.

The process of unscrambling the micronuclear DNA into the macronuclear genes is ostensibly aided by the presence of short pointer sequences present at the junctions between MDSs and IESs. More precisely, at the junction between the nth MDS and the adjacent IES following, it there is a sequence (of 9–13 bp in actin I and 6–19 bp in α telomere binding protein) that is repeated somewhere else in the gene—namely, at the junction between the $(n+1)^{st}$ MDS and the adjacent IES preceding it. After aligning two pointers, homologous recombination between them would then join MDSs n and $(n+1)$ in the correct order and eliminate one copy of the pointer.

This process can be viewed as a computation solved during gene unscrambling by homologous DNA recombinations. If we assume that the cell's biochemistry can identify those DNA segments that represent pointers, and if the pointer pairs were unique in the micronuclear DNA, then one could argue that the ciliate is solving the computational problem of sorting a permuted sequence in the correct order. However, the computational problem facing the cell is much more complex given that some pointer sequences occur more than 13 times in a single gene (e.g., DNA polymerase α in *S. lemnae*). Taking into account the multiplicities of each pointer (the raw number of occurrences of the sequence representing the pointer in the micronuclear sequence), the number of combinations the cell would need to explore in order to arrive at the correct solution could be greater than 14 trillion for DNA polymerase α in *S. lemnae* [2]. Clearly, even assuming a priori knowledge of the pointer sequences, blind searching of matching pointer pairs is not a realistic explanation of gene unscrambling.

Other factors such as knowledge of which DNA sequences represent MDSs, IESs, and pointers, together with geometric folding of the micronuclear DNA that brings corresponding pointers together, have been suggested as mechanisms of gene unscrambling [17]. Alternative or complementary information guiding unscrambling may be the presence of contexts that flank correct matching pointer pairs and that might be responsible for solving this seemingly difficult computational problem [17, 18]. The details of the gene rearrangement process are still elusive. In the following sections we explore formal models for the homologous recombinations that lead to gene unscrambling in ciliates and investigate their computational power.

CLASSIC COMPUTATIONAL MODELS

In order to study the computational power of the bio-operations underlying gene rearrangement in ciliates, we compare them with the existing models of computation. In this section, we introduce two classic models of computation, the finite automaton and the Turing machine. The finite automaton has very restricted computational power, but the Turing machine models the computing capability of a general-purpose computer.

The finite automaton is a mathematical model of a system with a finite number of internal states and with discrete inputs and outputs. The behavior of the system is dictated by a finite number of rules that, given a state of the system and an input, dictate what the next state of the system will be.

To formalize the notions of finite automaton and Turing machine, let us first introduce some notations. An alphabet Σ is a finite, nonempty set of symbols or letters. A sequence of letters from Σ is called a string (word) over Σ. (If $\Sigma = \{A, C, G, T\}$, a word over Σ can be interpreted as a linear DNA strand.) The words are denoted by lowercase letters such as u, v, α_i, x_{ij}. A word with 0 letters in it is called an empty word and is denoted by λ. The set of all possible words consisting of letters from Σ is denoted by Σ^*, and the set of all nonempty words by Σ^+.

A finite automaton is a construct $A = (S, \Sigma, P, s_0, F)$, where Σ is the input alphabet, S is a finite set of states, $s_0 \in S$ is a designated start state, $F \subseteq S$ is the set of final states, and $P \subseteq S \times \Sigma \longrightarrow S$ is a set of transition rules. A transition rule $sa \longrightarrow s'$, $s, s' \in S$, $a \in \Sigma$ says that, if the automaton is in state s and reads the input letter a, then it changes its state to the new state s' and continues scanning the input word. The language accepted by the automaton A is defined as:

$$L(A) = \{w \in \Sigma^* \mid s_0 w \Longrightarrow^* s_f, s_f \in F\};$$

in other words, the set of all input words that can take the automaton from the initial state to a final state by successive applications (denoted by $\Longrightarrow^*$) of the transition rules in P. In the hierarchy of computational models, finite automata are the weakest.

At the other end of the spectrum of computational power is the Turing machine (TM), the accepted formal model of what we call computation. In a Turing machine, a read/write head scans an infinite tape composed of discrete squares, one square at a time. The read/write head communicates with a control mechanism under which it can read the symbol in the current square or replace it by another. The read/write head is also able to move on the tape, one square at a time, to the right and to the left. The set of words that make a Turing machine finally halt is considered its language.

Formally [20], a rewriting system TM $= (S, \Sigma \cup \{\#\}, P)$ is called a Turing machine if and only if (iff):

(1) S and $\Sigma \cup \{\#\}$ (with $\# \notin \Sigma$ and $\Sigma \neq \emptyset$) are two disjoint alphabets referred to as the *state* and the *tape* alphabets.

(2) Elements s_0 and s_f of S and B of Σ are the *initial* and *final* state, and the *blank symbol*, respectively. Also, a subset T of Σ is specified and referred to as the *terminal* alphabet. It is assumed that T is not empty.

(3) The productions (rewriting rules) of P are of the forms

  (i) $s_i a \longrightarrow s_j b$ (overprint)
  (ii) $s_i ac \longrightarrow as_j c$ (move right)
  (iii) $s_i a\# \longrightarrow as_j B\#$ (move right and extend workspace)
  (iv) $cs_i a \longrightarrow s_j ca$ (move left)
  (v) $\#s_i a \longrightarrow \#s_j Ba$ (move left and extend workspace)
  (vi) $s_f\, a \longrightarrow s_f$
  (vii) $a\, s_f \longrightarrow s_f$

where s_i and s_j are in S, $s_i \neq s_f$, $s_j \neq s_f$, and a, b, c are in Σ. For each pair (s_i, a), where s_i and a are in the appropriate ranges, P either contains no productions (ii) and (iii) (respectively, iv and v) or else contains both (iii) and (ii) for every c (respectively contains both (v) and (iv) for every c). There is no pair (s_i, a) such that the word $s_i a$ is a subword of the left side in two productions of the forms i, iii, v.

A configuration of the TM is of the form $\#w_1 s_i w_2\#$, where $w_1 w_2$ represents the contents of the tape, #s are the boundary markers, and the position of the state symbol s_i indicates the position of the read/write head on the tape: if s_i is positioned at the left of a letter a, this indicates that the read/write head is placed over the cell containing a. The TM changes from one configuration to another according to its rules. For example, if the current configuration is $\#ws_i aw'\#$ and the TM has the rule $s_i a \longrightarrow s_j b$, this means that the read/write head positioned over the letter a will write b over it and change its state from s_i to s_j. The next configuration in the derivation will be thus $\#ws_j bw'\#$.

The TM halts with a word w iff there exists a derivation that, when started with the read/write head positioned at the beginning of w eventually reaches the final state (i.e. if $\#s_0 w\#$ derives $\#s_f\#$ by successive applications of the rewriting rules i–vii). The language $L(TM)$ accepted by TM consists of all words over the terminal alphabet T for which the TM halts. Note that TM is deterministic: at each step of the rewriting process, the application of at most one production is possible.

A FORMAL MODEL OF GENE REARRANGEMENT

In this section we describe several bio-operations we have studied as models of the homologous recombinations apparently underlying the process of gene unscrambling in ciliates [6–9]. We study these operations from a computational point of view. We namely summarize results in Kari and Kari [6] showing that, if the recognition of identical pointers is assumed to be sufficient to trigger recombination, the computational potential achieved is only that of finite automata.

We define circular words over Σ by declaring two words to be equivalent iff one is a cyclic permutation of the other. In other words, w is equivalent to w' iff they can be decomposed as $w = uv$ and $w' = vu$, respectively. Such a circular word $\bullet w$ refers to any of the circular permutations of the letters in w. Denote by $\Sigma^\bullet$ the set of all circular words over Σ.

DEFINITION 1   If $x \in \Sigma^+$ is a pointer, then the recombinations guided by x are defined as follows:

$$uxv + u'xv' \implies uxv' + u'xv \text{ (linear/linear)} \tag{1}$$

$$uxvxw \implies uxw + \bullet vx \text{ (linear/circular)} \tag{2}$$

$$\bullet uxv + \bullet u'xv' \implies \bullet uxv'u'xv \text{ (circular/circular).} \tag{3}$$

(See figures in Kari and Kari [6].) Note that all recombinations in definition 1 are reversible; the operations can be performed also in the opposite directions.

For example, operation (2) models the process of intramolecular recombination (Figure 10.2). After the pointer x finds its second occurrence in $uxvxw$, the molecule undergoes a strand exchange in x that leads to the formation of two new molecules: uxw and a circular DNA molecule $\bullet vx$. Intramolecular recombination accomplishes the deletion of either sequence vx or xv from the original molecule $uxwxv$ and the positioning of w immediately next to ux.

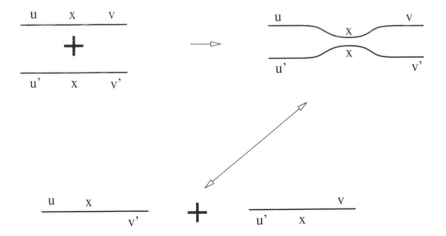

Figure 10.2 Linear/linear recombination.

This implies that operation 2 can be used to rearrange sequences in a DNA molecule, thus accomplishing gene unscrambling.

The above operations are similar to the the "splicing operation" introduced by Head [3] and circular splicing and mixed splicing [4, 14–16, 21]. It was subsequently shown that some of these models have the computational power of a universal Turing machine [1, 13, 22]. (See Head et al. [5] for a review.)

The process of gene unscrambling entails a series of successive or possibly simultaneous intra- and intermolecular homologous recombinations. This is followed by excision of all sequences $\tau_s y \tau_e$, where the sequence y is marked by the presence of telomere addition sequences τ_s for telomere "start" (at its 5' end), and τ_e for telomere "end" (at its 3' end). Thus, from a long sequence $u \tau_s y \tau_e v$, this step retains only $\tau_s y \tau_e$ in the macronucleus. Last, the enzyme telomerase extends the length of the telomeric sequences (usually double-stranded *TTTTGGGG* repeats in these organisms) from τ_s and τ_e to protect the ends of the DNA molecule.

We now make the assumption that, either by a structural alignment of the DNA or by other biochemical factors, the cell decides which sequences are non–protein-coding (IESs) and which are ultimately protein coding (MDSs), as well as which are the pointers x. Such biological shortcuts are presumably essential to bring into proximity the pointers x. Each of the n MDSs, denoted primarily by α_i, $1 \leq i \leq n$, is flanked by the pointers $x_{i-1,i}$ and $x_{i,i+1}$. Each pointer points to the MDS that should precede or follow α_i in the final sequence. The only exceptions are α_1, which is preceded by τ_s, and α_n, which is followed by τ_e in the input string or micronuclear molecule. Note that, although present generally once in the final macronuclear copy, each $x_{i,i+1}$ occurs at least twice in the micronuclear copy: once after α_i and once before α_{i+1}.

We denote by ϵ_k an internal sequence that is eliminated; ϵ_k does not occur in the final sequence. Thus, since unscrambling leaves one copy of each $x_{i,i+1}$ between α_i and α_{i+1}, an IES is nondeterministically either $\epsilon_k x_{i,i+1}$ or $x_{i-1,i} \epsilon_k$, depending on which pointer $x_{i,i+1}$ is eliminated. Similarly, an MDS is technically either $\alpha_i x_{i+1}$ or $x_{i-1,i} \alpha_i$. For this model, either choice is equivalent.

The following example (from Landweber and Kari [8]) models unscrambling of a micronuclear gene that contains MDSs in the scrambled order 2-4-1-3 using only the operation of linear/circular recombination:

$$\{u \; x_{12} \; \alpha_2 \; x_{23} \; \epsilon_1 \; x_{34} \; \alpha_4 \; \tau_e \; \epsilon_2 \; \tau_s \; \alpha_1 \; x_{12} \; \epsilon_3 \; x_{23} \; \alpha_3 \; x_{34} \; v\} \Longrightarrow$$

$$\{u \; x_{12} \; \epsilon_3 \; x_{23} \; \alpha_3 \; x_{34} \; v \;, \quad \bullet \alpha_2 \; x_{23} \; \epsilon_1 \; x_{34} \; \alpha_4 \; \tau_e \; \epsilon_2 \; \tau_s \; \alpha_1 \; x_{12}\}$$

$$= \{u \; x_{12} \; \epsilon_3 \; x_{23} \; \alpha_3 \; x_{34} \; v, \quad \bullet \epsilon_1 \; x_{34} \; \alpha_4 \; \tau_e \; \epsilon_2 \; \tau_s \; \alpha_1 \; x_{12} \; \alpha_2 \; x_{23}\} \Longrightarrow$$

$$\{u \; x_{12} \; \epsilon_3 \; x_{23} \; \epsilon_1 \; x_{34} \; \alpha_4 \; \tau_e \; \epsilon_2 \; \tau_s \; \alpha_1 \; x_{12} \; \alpha_2 \; x_{23} \; \alpha_3 \; x_{34} \; v\} \Longrightarrow$$

$$\{u \; x_{12} \; \epsilon_3 \; x_{23} \; \epsilon_1 \; x_{34} \; v, \quad \bullet\alpha_4 \; \tau_e \; \epsilon_2 \; \tau_s \; \alpha_1 \; x_{12} \; \alpha_2 \; x_{23} \; \alpha_3 \; x_{34}\}$$

$$= \{u \; x_{12} \; \epsilon_3 \; x_{23} \; \epsilon_1 \; x_{34} \; v \;, \quad \bullet\tau_s \; \alpha_1 \; x_{12} \; \alpha_2 \; x_{23} \; \alpha_3 \; x_{34} \; \alpha_4 \; \tau_e \; \epsilon_2\} \Longrightarrow$$

$$\{\tau_s \; \alpha_1 \; x_{12} \; \alpha_2 \; x_{23} \; \alpha_3 \; x_{34} \; \alpha_4 \; \tau_e \;, \quad \epsilon_2 \;, \quad u \; x_{12} \; \epsilon_3 \; x_{23} \; \epsilon_1 \; x_{34} \; v\}.$$

The case in which all types of recombinations (linear/linear, linear/circular (Fig. 10.3), circular/circular (Fig. 10.4)) can occur, without restrictions, has been studied [6]. This study complements results obtained on linear splicing, circular splicing, self-splicing, and mixed splicing [5, 14–16, 21]. However, while theorem 2 may follow from a result in Head et al. [5] on the closure of an Abstract Family of Languages (AFL) under all splicings, theorem 1 characterizes the language, $L(R)$, of an arbitrary, context-free recombination system with a possibly infinite set of pointers and arbitrary axiom sets.

The intuitive image of context-free recombinations is that one can view strings as cables or extension cords with different types of plugs. Given a set of pointers J, each $x \in J$ defines one type of plug. Strings, both linear and circular, can then be viewed as consisting of elementary cables that only have plugs at their extremities. (A circular strand consists of elementary cables connected to form a loop.) A recombination step amounts to the following operations: take two connections using identical plugs (the connections can be in two different cables or in the same cable); unplug them; cross-plug to form new cables. We will assume, without loss of generality, that all sets of plugs J are subword-free [6].

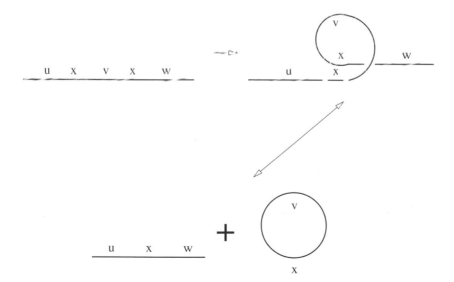

Figure 10.3 Linear/circular recombination.

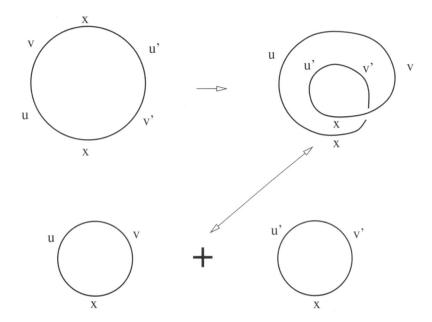

Figure 10.4 Circular/circular recombination.

DEFINITION 2   Let $J \subseteq \Sigma^+$ be a set of plugs. We define the set of elementary cables (respectively, left elementary cables and right elementary cables) with plugs in J as

$$E_J = (J\Sigma^+ \cap \Sigma^+ J) \setminus \Sigma^+ J\Sigma^+,$$

$$L_J = \Sigma^* J \setminus \Sigma^* J\Sigma^+,$$

$$R_J = J\Sigma^* \setminus \Sigma^+ J\Sigma^*.$$

Note that an elementary cable in E_J is of the form $z_1 u = v z_2$, where $z_1, z_2 \in J$ are plugs. In other words, an elementary cable starts with a plug, ends with a plug, and contains no other plugs as subwords. The start and end plug can overlap.

A left elementary cable is of the form wz, where $z \in J$ is a plug and wz does not contain any other plug as a subword. In other words, if we scan wz from left to right, z is the first plug we encounter. Analogously, a right elementary cable is of the form zw, where $z \in J$ is a plug and wz does not contain any other plug as a subword.

DEFINITION 3   For a set of plugs $J \subseteq \Sigma^+$ and a linear word $w \in \Sigma^+$, the set of elementary cables with plugs in J occurring in w is defined as

$$E_J(w) = E_J \cap Sub(w),$$

while the set of left and right elementary cables occurring in w is

$$L_J(w) = L_J \cap Pref(w)$$

$$R_J(w) = R_J \cap Suff(w),$$

respectively, where $Pref(w)$, $Suff(w)$, $Sub(w)$ denote the set of all prefixes, suffixes and respectively subwords of w.

One can prove that recombination of cables does not produce additional elementary cables [6]. In other words, the set of the elementary cables of the result strings equals the set of elementary cables of the strings entering recombination.

A *context-free recombination system* is a construct whereby we are given a starting set of sequences and a list of pointers (plugs). New strings may be formed by recombinations among the existing strands: if given pointers are present, recombinations are performed as defined. Recombinations are context-free (i.e., they are not dependent on the context in which the pointers appear). The language of the system is defined as the set of all strands that can be thus obtained by repeated recombinations starting from the initial set.

DEFINITION 4   A context-free recombination system is a triple

$$R = (\Sigma, J, A),$$

where Σ is an alphabet and $J \subseteq \Sigma^+$ is a set of plugs, while $A \subseteq \Sigma^+ \cup \Sigma^\bullet$ is the set of axioms of the system.

The theorem below shows that a context-free recombination system characterized by a set of plugs J and a set of axioms A has the following property: any cable that consists of elementary cables plugged together after each other and that is either linear or circular can be obtained from the axioms using cross-plugging. Conversely, no other types of cables can be obtained from the axioms.

THEOREM 1   Let $R = (\Sigma, J, A)$ be a context-free recombination system. Then $L(R) = X$ where $X = \{w \in \Sigma^* \cup \Sigma^\bullet|$ either $E_J(w) = L_J(w) = R_J(w) = \emptyset$ and $w \in A$ or $E_J(w), L_J(w), R_J(w)$ are not all empty and $E_J(w) \subseteq E_J(A)$, $L_J(w) \subseteq L_J(A), L_J(w) \subseteq L_J(A)\}$.

The theorem above will lead to the conclusion of this section provided we show that the language X is regular being accepted by a finite automaton. As X contains both linear and circular words, we have to first define the notion of acceptance of circular words by a finite automaton.

DEFINITION 5   Given a finite automaton $\mathcal{A}$, the circular language accepted by $\mathcal{A}$, denoted by $L(\mathcal{A})^\bullet$, is defined as the set of all words $\bullet w$ such that $\mathcal{A}$ has a cycle labeled by w.

The linear/circular language accepted by a finite automaton $\mathcal{A}$ is defined as $L(\mathcal{A}) \cup L(\mathcal{A})^{\bullet}$, where $L(\mathcal{A})$ is the linear language accepted by the automaton $\mathcal{A}$ defined in the usual way.

DEFINITION 6   A linear/circular language $L \subseteq \Sigma^* \cup \Sigma^{\bullet}$ is called *regular* if there exists a finite automaton $\mathcal{A}$ that accepts the linear and circular parts of L (i.e., that accepts $L \cap \Sigma^*$ and $L \cap \Sigma^{\bullet}$).

We can now formulate the main result presented in this section.

THEOREM 2   Let $J \subseteq \Sigma^*$ be a set of plugs and let $A \subseteq \Sigma^* \cup \Sigma^{\bullet}$ be a finite axiom set. Then the set X defined as in theorem 1 equals the linear/circular language accepted by a finite automaton $\mathcal{A}$ and is therefore regular.

Theorem 2 shows that the rewriting systems based on context-free recombinations are computationally weak, having only the power of finite automata. This is one more indicator that, most probably, the presence of pointers alone does not provide all the information needed for accurate splicing during gene rearrangement.

GUIDED RECOMBINATION SYSTEMS

As seen previously, the estimated running time of a pointer-search-and-match algorithm simulating gene rearrangement is prohibitively high. The previous section showed that the computational power of a formal computational model based on such context-free recombinations is very low—namely, that of finite automata. These and other biological arguments point to the fact that this model should be further refined to accurately reflect the biological reality [17]. We have introduced the additional assumption that homologous recombination is influenced by the presence of certain *guiding contexts* flanking the pointers present at the MDS-IES junctions [9]. The observed dependence on the old macronuclear sequence for correct IES removal in the distantly related ciliate *Paramecium* suggests that this is the case [11]. This restriction captures the fact that the pointers do not contain all the information for accurate splicing during gene unscrambling. In particular, we defined the notion of a *guided recombination system* based on operation 2 and proved that such systems have the computational power of a TM, the most widely used theoretical model of electronic computers [9].

We defined the contexts that restrict the use of recombinations by a *splicing scheme* [3, 4, 9] a pair $(\Sigma, \sim)$ where Σ is the alphabet and $\sim$, the pairing relation of the scheme, is a binary relation between triplets of nonempty words satisfying the following condition: if $(p, x, q) \sim (p', y, q')$, then $x = y$.

In the splicing scheme $(\Sigma, \sim)$, pairs $(p, x, q) \sim (p', x, q')$ now define the contexts necessary for a recombination between pointers x. Then we define *contextual intramolecular recombination* as

$$\{uxwxv\} \implies \{uxv, \bullet wx\},$$

where $u = u'p, w = qw' = w''p', v = q'v'$. This constrains intramolecular recombination within $uxwxv$ to occur only if the restrictions of the splicing scheme concerning x are fulfilled (i.e., the first occurrence of x is preceded by p and followed by q, and its second occurrence is preceded by p' and followed by q').

Similarly, if $(p, x, q) \sim (p', x, q')$, then we define *contextual intermolecular recombination* as

$$\{uxv, \bullet wx\} \implies \{uxwxv\},$$

where $u = u'p, v = qv', w = w'p' = q'w''$. Informally, intermolecular recombination between the linear strand uxv and the circular strand $\bullet wx$ may take place only if the occurrence of x in the linear strand is flanked by p and q and its occurrence in the circular strand is flanked by p' and q'. Note that sequences p, x, q, p', q' are nonempty and that both contextual intra- and intermolecular recombinations are reversible by introducing pairs $(p, x, q') \sim (p', x, q)$ in $\sim$.

The operations defined in the preceding section are particular cases of contextual recombinations, where all the contexts are empty [i.e, $(\lambda, x, \lambda) \sim (\lambda, x, \lambda)$ for all $x \in \Sigma^+$]. This would correspond to the case where recombination may occur between every two pointers, regardless of their contexts.

DEFINITION 7   A guided recombination system is a construct $R = (\Sigma, \sim, A)$ where $(\Sigma, \sim)$ is a splicing scheme, and $A \in \Sigma^+$ is a linear string called the *axiom*.

Those strands which, by repeated contextual recombinations with initial and intermediate strands eventually produce the axiom, form the language of the guided recombination system, $L_a^k(R)$. $L_a^k(R)$ thus denotes the multiset of words $w \in \Sigma^*$ with the property that, if present initially in at least k copies, are able to produce the axiom A by a series of contextual recombinations. (A multiset is a set where to each element is associated a multiplicity. Operations applied to elements of a multiset change their multiplicities. After an operation, the multiplicities of the operation inputs decrease by one, while the multiplicity of the operation output increases by one. The need of multisets for modeling is justified in Landweber and Kari [9].)

THEOREM 3   Let L be a language over T^* accepted by TM $= (S, \Sigma \cup \{\#\}, P)$. Then there exist an alphabet Σ', a sequence $\pi \in \Sigma'^*$, depending on L, and a guided recombination system, R, such that a word w over T^* is in L if and only if $\#^6 s_0 w \#^6 \pi$ belongs to $L_a^k(R)$ for some $k \geq 1$.

The idea of the proof is as follows. Consider that the rules of P are ordered in an arbitrary fashion and numbered. Thus, if TM has m rules, a rule is of the form $i : u_i \longrightarrow v_i$ where $1 \leq i \leq m$.

We construct a guided recombination system $R = (\Sigma', \sim, A)$ and a sequence $\pi \in \Sigma'^*$ with the required properties. The alphabet is $\Sigma' = S \cup \Sigma \cup \{\#\} \cup \{\$_i \mid 0 \leq i \leq m + 1\}$. The axiom, i.e., the target string to be achieved at the end of the computation, consists of the final state of the TM bounded by markers:

$$A = \#^{n+2} s_f \; \#^{n+2} \$_0 \$_1 \ldots \$_m \$_{m+1},$$

where n is the maximum length of the left-side or right-side words of any of the rules of the TM.

The sequence π consists of the catenation of the right-hand sides of the TM rules bounded by markers, as follows:

$$\pi = \$_0 \; \$_1 e_1 v_1 f_1 \$_1 \; \$_2 e_2 v_2 f_2 \$_2 \ldots \$_m e_m v_m f_m \$_m \; \$_{m+1},$$

where $i : u_i \longrightarrow v_i$, $1 \leq i \leq m + 1$ are the rules of TM and $e_i, v_i \in \Sigma \cup \{\#\}$.

If a word $w \in T^*$ is accepted by the TM, a computation starts then from a strand of the form $\#^{n+2} s_0 w \#^{n+2} \pi$, where we will refer to the subsequence starting with $\$_0$ as the "program" and to the subsequence at the left of $\$_0$ as the "data."

We construct the relation $\sim$ defining the contexts guiding the computations so that (1) the right-hand sides of rules of TM can be excised from the program as circular strands which then interact with the data; and (2) When the left-hand side of a TM rule appears in the data, the application of the rule can be simulated by the insertion of the circular strand encoding the right-hand side, followed by the deletion of the left-hand side.

With the help of these contexts, we can prove theorem 3—namely, that we can simulate the computation of any given TM by a series of contextual inter- and intramolecular recombinations.

Theorem 3 also implies that if a word $w \in T^*$ is in $L(TM)$, then $\#^6 s_0 w \#^6 \pi$ belongs to $L_a^k(R)$ for some k, and therefore it belongs to $L_a^i(R)$ for any $i \geq k$. This means that, to simulate a computation of the TM on w, any sufficiently large number of copies of the initial strand will do. The assumption that sufficiently many copies of the input strand are present at the beginning of the computation is in accordance with the fact that there are multiple copies of each strand available during the (polytene chromosome) stage where unscrambling occurs. Note that the preceding result is valid even if we allow interactions of operation 3 between circular strands or within a circular strand.

The proof that a guided recombination system can simulate a TM and thus any computation suggests that a functional macronuclear gene can be viewed as the output of a computation performed on the micronuclear sequence.

CONCLUSIONS

We have described a model for the process of gene rearrangement in spirotrich ciliates. Although the model is consistent with our limited knowledge of this biological process, it awaits rigorous testing by the tools of molecular genetics. The model in its present form is capable of universal computation. This hints at future directions and the use of ciliates as model systems for exploring cellular computation.

Acknowledgments   This research was partially supported by the Natural Sciences and Engineering Research Council of Canada grant R2824AO1 to L.K. and National Institute of General Medical Sciences grant GM59708 to L.F.L. We thank Mark Daley for comments and Jeremy Newton-Smith with help in preparing the manuscript.

References

[1] E. Csuhaj-Varju, R. Freund, L. Kari, and G. Paun. DNA computing based on splicing: universality results. In L. Hunter and T. Klein, editors, *Proceedings of 1st Pacific Symposium on Biocomputing*, pages 179–190. World Scientific Publishing, Singapore, 1996.

[2] M. Daley. Complexity of gene unscrambling, 2002. Unpublished manuscript.

[3] T. Head. Formal language theory and DNA: an analysis of the generative capacity of specific recombinant behaviors. *Bull. Math. Biol.*, 49:737–759, 1987.

[4] T. Head. Splicing schemes and DNA. In G. Rozenberg and A. Salomaa, editors, *Lindenmayer Systems*, pages 371–383. Springer-Verlag, Berlin, 1991.

[5] T. Head and G. Paun, and D. Pixton. Language theory and molecular genetics. In G. Rozenberg and A. Salomaa, editors, *Handbook of Formal Languages*, vol. 2, pages 295–358. Springer-Verlag, Berlin, 1997.

[6] J. Kari and L. Kari. Context-free recombinations. In C. Martin-Vide and V. Mitrana, editors, *Where Mathematics, Computer Science, Linguistics and Biology Meet*, pages 361–375. Kluwer, Dordrecht, The Netherlands, 2001.

[7] L. Kari, J. Kari, and L. Landweber. Reversible molecular computation in ciliates. In J. Karhumaki, H. Maurer, G. Paun, and G. Rozenberg, editors, *Jewels are Forever*, pages 353–363. Springer-Verlag, Berlin, 1999.

[8] L. F. Landweber and L. Kari. The evolution of cellular computing: nature's solution to a computational problem. *BioSystems*, 52(1–3):3–13, 1999.

[9] L. F. Landweber and L. Kari. Universal molecular computation in ciliates. In L. Landweber and E. Winfree, editors, *Evolution as Computation*. Springer-Verlag, Berlin, 2002.

[10] L. F. Landweber, T. Kuo, and E. Curtis. Evolution and assembly of an extremely scrambled gene. *Proc. Natl. Acad. Sci.*, 97(7):3298–3303, 2000.

[11] E. Meyer and S. Duharcourt. Epigenetic programming of developmental genome rearrangements in ciliates. *Cell*, 87:9–12, 1996.

[12] J. L. Mitcham, A. J. Lynn, and D. M. Prescott. Analysis of a scrambled gene: the gene encoding α-*telomere-binding* protein in *Oxytricha nova*. *Genes Devel.*, 6:788–800, 1992.

[13] G. Paun. On the power of the splicing operation. *Int. J. Comp. Math*, 59:27–35, 1995.

[14] D. Pixton. Linear and circular splicing systems. In *Proceedings of the First International Symposium on Intelligence in Neural and Biological Systems*, pages 181–188. IEEE Computer Society Press, Los Alamos, NM, 1995.

[15] D. Pixton. Regularity of splicing languages. *Discrete Appl. Math.*, 69(1–2):99–122, 1996.

[16] D. Pixton. Splicing in abstract families of languages. *Theor. Computer Sci.*, 234:135–166, 2000.

[17] D. M. Prescott. The DNA of ciliated protozoa. *Microbiol. Rev.*, 58(2):233–267, 1994.

[18] D. M. Prescott. Genome gymnastics: unique modes of dna evolution and processing in ciliates. *Nature Rev. Genet.*, 1(3):191–198, 2000.

[19] D. M. Prescott and A. M. Greslin. The scrambled actin i gene in the micronucleus of *Oxytricha nova*. *Devel. Genet.*, 13:66–74, 1992.

[20] A. Salomaa. *Formal Languages*. Academic Press, New York, 1973.

[21] R. Siromoney, K.G. Subramanian, and V. Rajkumar Dare. *Circular DNA and Splicing Systems*, pages 260–273. Springer-Verlag, Berlin, 1992.

[22] T. Yokomori, S. Kobayashi, and C. Ferretti. On the power of Circular splicing systems and DNA computability. In *Proceedings of IEEE International Conference on Evolutionary Computation*, pages 219–224, IEEE Press, Piscataway, NJ, 1997.

Index